Legumes in livestock and fish nutrition

Legumes in livestock and fish nutrition

Ingredients, feed value and feeding recommendations

Edited by

Gerhard Bellof
Mechthild Freitag

BRILL | WAGENINGEN ACADEMIC

The Library of Congress Cataloging-in-Publication Data is available online at https://catalog.loc.gov
LC record available at https://lccn.loc.gov/2024952269

Typeface for the Latin, Greek, and Cyrillic scripts: "Brill". See and download: brill.com/brill-typeface.

ISBN 978-90-04-69503-0 (hardback)
ISBN 978-90-04-69504-7 (e-book)
DOI 10.3920/9789004695047

This book is printed on acid-free paper and produced in a sustainable manner.

Contents

Figures and tables

Figures

Tables

Abbreviations

AA	Amino Acid
ADC	Apparent Digestibility Coefficient
ADF	Acid Detergent Fibre
ADF_{om}	Acid Detergent Fibre in organic matter
aNDFom	Neutral Detergent Fiber in organic matter after amylase treatment
ANF	Antinutritive Factor
ASH	Ash
BE	Base excess
BW	Body weight
CF	Crude Fiber
CL	Crude Lipid / Ether Extract
CNCPS	Cornell Net Carbohydrate and Protein System
CP	Crude Protein
CPD	Digestible Crude Protein
DC	Digestibility Coefficient
DCAB	Dietary Cation-Anion Balance
DCNDF	Digestibility Coefficient of NDF
DCOM	Digestibility Coefficient of Organic Matter
DM	Dry Matter
EAA	Essential Amino Acid
ECM	Energy-Corrected Milk
FB	Faba Bean
FBI	Faba Bean Isolate
FBM	Faba Bean Meal
FCR	Feed Conversion Ratio
FECM	Fat and Protein Corrected Milk
FM	Fish Meal
NDF	Neutral Detergent Fibre
NFE	Nitrogen Free Extract
NSP	Non-Starch Polysaccarides
nXP	Crude Protein at the Duodenum
OM	Organic Matter
OMD	Digestible Organic Matter
pcdAA	pre-caecal digestible Amino Acid
PM	Pea Meal
PP	Pea Protein
PPC	Pea Protein Concentrate
PPI	Pea Protein Isolate
RUP	Rumen Undegraded Protein

SBM	Soybean Meal
SEM	Standard Error of the Mean
SGR	Specific Growth Rate
SPC	Soy Protein Concentrate
SPI	Soy Protein Isolate
STA	Starch
SUG	Sugar
TIA	Trypsin Inhibitor Activity
TMR	Total Mixed Ration
UDP	Undegraded Dietary Protein
V/C	Vicine and Convicine
VFA	Volatile Fatty

Contributors

Prof. Dr. Gerhard Bellof
Professur Tierernährung / Professorship of Animal Nutrition
Hochschule Weihenstephan-Triesdorf
Fakultät Nachhaltige Agrar- und Energiesysteme
85350 Freising, Germany
gerhard.bellof@hswt.de

Prof. Dr. Mechthild Freitag
Professur für Physiologie und Ernährung / Professorship of Physiology and Nutrition
Fachhochschule Südwestfalen
Fachbereich Agrarwirtschaft
59494 Soest, Germany
freitag.mechthild@fh-swf.de

Dr. Manuela Specht
Wissenschaftliche Mitarbeiterin für Produktionsmanagement der Öl- und Proteinpflanzen, Tierernährung und Humanernährung / Research associate for Oil and protein plants, animal nutrition and human nutrition
Union zur Förderung von Oel- und Proteinpflanzen e.v. (UFOP)
10117 Berlin, Germany
M.Specht@ufop.de

Prof. Dr. Petra Weindl
Professur Nachhaltige Tierernährung und Futtermittelkunde / Professorship of Sustainable Animal Nutrition and Feed Science
Hochschule Weihenstephan-Triesdorf
Fakultät Nachhaltige Agrar- und Energiesysteme
85354 Freising, Germany
petra.weindl@hswt.de

Dr. Stephan Hartmann
Leiter der Arbeitsgruppe Züchtungsforschung bei Futterpflanzen und Pflanzenbausysteme im Futterbau / Head of Breeding research in forage plants and plant cultivation systems in forage production
Institut für Pflanzenbau und Pflanzenzüchtung
Bayerische Landesanstalt für Landwirtschaft
85356 Freising, Germany
Stephan.Hartmann@lfl.bayern.de

Dr. Monika Weiß
Leitung Themenbereich Futter und nachhaltige Ressourcen / Head of the topics feed and sustainable resources
Alfred-Wegener-Institut
Helmholtz-Zentrum
für Polar- und Meeresforschung
27570 Bremerhaven, Germany
Monika.Weiss@awi.de

Dr. Vanessa Fuchs
Wissenschaftliche Mitarbeiterin für Tierernährung und nachhaltige Futtermittel in der Aquakultur / Research associate for animal nutrition and sustainable feed in aquaculture
Alfred-Wegener-Institut
Helmholtz-Zentrum
für Polar- und Meeresforschung
27570 Bremerhaven, Germany
Vanessa.Fuchs@awi.de

Dr. Sinem Zeytin
Wissenschaftliche Mitarbeiterin für Tierernährung und nachhaltige Futtermittel in der Aquakultur / Research associate for animal nutrition and sustainable feed in aquaculture
Alfred-Wegener-Institut
Helmholtz-Zentrum
für Polar- und Meeresforschung
27570 Bremerhaven, Germany
sinem.zeytin@awi.de

CHAPTER 1

Legumes – classification and differentiation

Gerhard Bellof, Manuela Specht and Mechthild Freitag

Legumes are a very diverse plant family and botanically belong to the order *Fabales*. The species include perennial as well as annual and biennial herbaceous plants and also woody plants. For nutritional purposes, the important legume species are annual herbaceous plants that are classified as grain legumes. Green or forage legumes are perennial flowering plants in the legume family *Fabaceae*.

As agricultural protein crops, grain legumes are used either directly for human consumption or as feed for livestock. Depending on the composition of the seed, the various grain legume species are used in animal diets primarily as a pure protein supplement or as both a protein and energy supplement in the feed mixtures or rations.

Reducing the use of feedstuffs that compete with humans for food is of central importance for greater sustainability in livestock farming (Eisler et al., 2014). Nowadays, the human edible fraction (HEF) is taken into account when evaluating animal feed. The digestive system of monogastric animals is similar to that of humans. In terms of nutrient composition, the nutritional requirements for pigs and poultry are therefore comparable to those of humans. For example, high protein soybean meal (48% crude protein), used in large proportions in pig and poultry feed worldwide, has a human edible fraction of 90%. The use of small-grain legumes – that are not part of human diets – as a source of protein and amino acids for pigs and poultry could therefore reduce the HEF in pork and poultry feed.

In Germany, the Federal Ministry of Food and Agriculture (BMEL) initiated a Protein Crop Strategy at the end of 2012 with the aim of reducing dependence on imported protein. The strategy is intended to support an expansion in the cultivation and utilisation of legumes – both in terms of supply and demand. The aim is to strengthen the competitiveness of domestically produced legumes. The Protein Crop Strategy is geared equally towards conventional and organic cultivation (BLE, 2024). The BMEL Arable Farming Strategy from 2021 sets a target of 10% legume cultivation on German arable land by 2030 (BMEL, 2021). Furthermore, the Union zur Förderung von Oel- und Proteinpflanzen e.V. (Union for the Promotion of Oil and Protein Plants, UFOP) – the association representing the interests of oil and protein crop producing and processing stakeholders in Germany – has identified the potential for the combined cultivation of rapeseed and legumes on 20% of German arable land as part of the development of its 10 + 10 strategy (Bellof *et al.*, 2021). Measures to promote domestic protein crop production are also being taken in other EU member states, implemented as part of Common Agricultural

 | DOI:10.3920/9789004695047_002

Policy strategy plans from 2023. The EU Commission has compiled a corresponding overview – the promotion of protein crop production is possible both as a coupled payment or eco-scheme under Pillar 1 and as an agri-environmental measure under Pillar 2 (EU Commission, 2023). In May 2024, the EU Commission published the final report of a "Study on feeding strategies to diversify the protein sources used in different livestock production systems in the EU". The study recommends a package of measures and a comprehensive research strategy to expand the production and use of grain legumes (EU Commission, 2024). In 2023, a working group of representatives from research institutions in various European countries simulated various scenarios with the result that grain legumes can play a significant role in the desired shift towards more plant-based foods. However, in order to increase the degree of self-sufficiency with plant protein, both an expansion of the area and an increase in the yields of grain peas, faba beans and soybeans are necessary. If it is possible to replace animal proteins in the diet with plant proteins within European acceptance limits, this could lead to an overall reduction in arable land requirements of 4–8% and a reduction in greenhouse gas emissions of 7–22% compared to current meat production (Van Loon *et al.*, 2023).

Due to their high protein content and protein quality, the cultivation of soybeans is of particular interest. Cultivation is currently limited to middle and southern Europe due to the lack of climate-adapted varieties. Nevertheless, the area under cultivation in Germany has increased from 3,500 in 2010 to 41,700 ha in 2024 (Statista, 2024). In the future, soybeans could be the first choice among legumes (Zimmer and Böttcher, 2021).

References

Bellof G., Specht M., Hotte S., Kleimeier C., Lenz H., Peters J., Richardt W., Sass O., Sieling K., Weber M., 2022: Szenarien für den Anbau und die Verwertung von Leguminosen und Körnerraps in Deutschland im Jahr 2030. Berichte über Landwirtschaft, 100, 1–39. https://doi.org/10.12767/buel.v100i2.417.

BLE, 2024: https://www.ble.de/DE/Projektfoerderung/Foerderungen-Auftraege/Eiweisspflanzenstrategie/eiweisspflanzenstrategie_node.html. Retrieved July 3, 2024.

BMEL, 2021: Ackerbaustrategie 2035. https://www.bmel.de/SharedDocs/Downloads/DE/Broschueren/ackerbaustrategie2035.pdf?__blob=publicationFile&v=8. Retrieved July 3, 2024.

Eisler M.C., Lee M.R.F., Tarlton J.F., Martin G.B., Beddington J., Dungait J.A.J., Greathead H., Liu J., Mathew S., Miller H., Misselbrook T., Murray P., Vinod V.K., van Saun R., Winter M., 2014: Agriculture: steps to sustainable livestock. Nature, 507, 32–34. doi:10.1038/507032a.

EU-Kommission, 2023: Approved 28 CAP strategic plans (2023–2027) summary overview for 27 member states – facts and figures. https://agriculture.ec.europa.eu/document/download/7b3a0485-c335-4e1b-a53a-9fe3733ca48f_en?filename=approved-28-cap-strategic-plans-2023-27.pdf&prefLang=de. Retrieved July 3, 2024.

EU-Kommission, 2024: Study on feeding strategies to diversity protein sources used in different livestock production systems in the EU. https://op.europa.eu/en/publication-detail/-/publication/7a284525-198c-11ef-a251-01aa75ed71a1. Retrieved July 3, 2024.

Statista, 2024: https://de.statista.com/statistik/daten/studie/1172777/umfrage/anbauflaeche-sojabohnen-deutschland.

Zimmer Y., Böttcher T., 2021: Mit Sojaanbau profitabel Fruchtfolgen erweitern? Thünen Working Paper 169.

Van Loon M.P., Alimagham S., Pronk A., Fodor N., Ion V., Kryvoshein O., Kryvobok O., Marrou H., Mihail R., Mínguez M.I., Pulina A., Reckling M., Rittler L., Roggero P.P., Stoddard F.L., Topp C.F.E., van der Wel J., Watson C., van Ittersum M.K., 2023: Grain legume production in Europe for food, feed and meat-substitution. Global food security, 39, 100723. doi: 10.1016/j.gfs.2023.100723.

CHAPTER 2

Grain legumes

2.1 Importance and cultivation of grain legumes

Manuela Specht

2.1.1 *Importance*

Grain legumes as agriculturally utilised protein plants are used either directly for human nutrition or livestock nutrition. Depending on the composition of the seed, the various grain legume species are used as animal feed primarily as a protein supplement or as both a protein and energy supplement in the concentrate mixtures or rations.

The cultivation area of faba bean (*Vicia faba L.*), grain pea (*Pisum sativum L.*), blue or narrow-leaved and white sweet lupin (*Lupinus angustifolius L.* and *Lupinus albus L.*) and soybean (*Glycine max L.*) in Germany has reached a total of 249,200 ha for the 2023 harvest. Please refer to figure 2.1.1-1 for the breakdown between the above-mentioned crops and the development of cultivation over the last 25 years.

In addition, 27,200 ha of other grain legumes and mixed crops were cultivated for grain production in Germany in 2023. This means that grain legumes accounted for approx. 2.4% of the German arable land. Together with small-grain legumes or green legumes (red clover, lucerne, mixtures of 80% or more small-grain legumes), they accounted for approx. 5.4% of arable land in Germany in 2023 (Destatis, 2024).

In the EU, not all grain legumes are cultivated equally in all member states. For the 2023 harvest, there were cultivation priorities for grain peas in Spain, France and Germany, for faba beans in France, Germany and Italy, and for sweet lupins in Poland, Germany and Greece. The former EU member state Great Britain is the largest producer of faba beans in Europe. The main soybean producing countries in the EU are Italy, France, Romania and Austria (UFOP, 2024; according to EU Commission, 2024). Another important soybean producing country in Europe is Ukraine, which exports a very large proportion of its soybean harvest to the EU.

Legumes form a symbiosis with their roots with nitrogen-fixing bacteria, so-called rhiziobia. For this purpose, root nodules are formed. When active, they are reddish to red in colour inside. With the help of this symbiosis, the legume can use atmospheric nitrogen from the soil air provided via fixation for growth and yield formation. This makes the legume independent on the supply of mineral nitrogen. The rhizobia in turn receive metabolic products from the photosynthetic performance of the legume plant. If the arable soil on which legumes are grown is rich in readily available mineral nitrogen, nodule formation is inhibited or does not occur at all, even if the appropriate rhizobia are generally present in the soil. In Germany,

 | DOI:10.3920/9789004695047_003

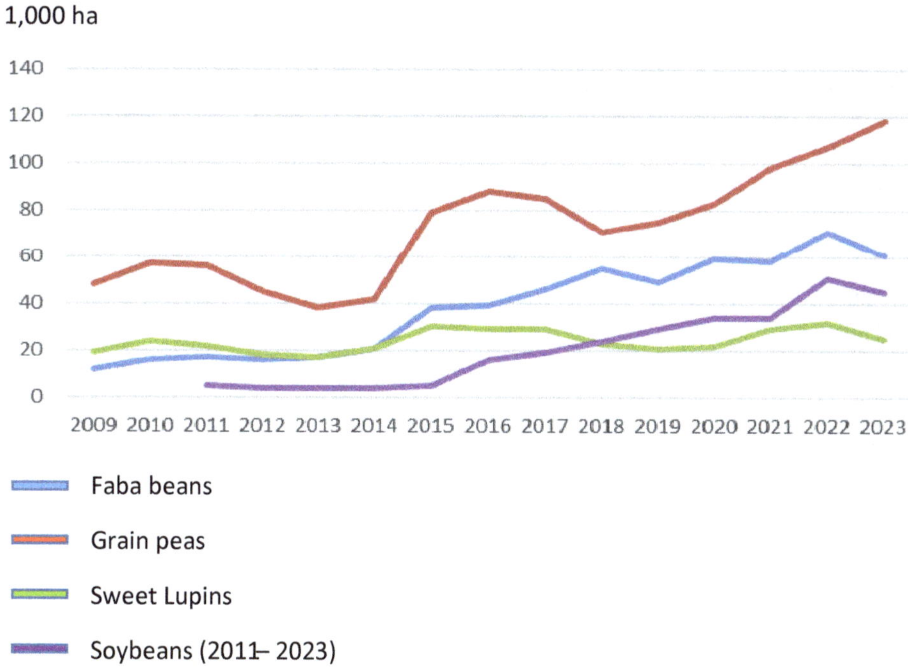

FIGURE 2.1.1-1 Crop development for faba bean, grain pea, blue or narrow-leaved and white sweet lupin and soybean 2009 to 2023 in Germany (in 1,000 ha)
DESTATIS (2010-2023)

rhizobia for faba beans and grain peas are naturally widespread in arable soils, so that they do not need to be inoculated for the cultivation of the above-mentioned crops. For sweet lupins, it is recommended to inoculate rhizobia during the first cultivation or after a longer break in cultivation. For soybeans, regular inoculation is recommended for each cultivation, as these rhizobia are not native. As a rule, seeds are inoculated with commercially available products immediately before sowing, whereby the seeds should not be exposed to high temperatures and intense sunlight afterwards. In some cases, preparations for application to the soil are also available. Due to their ability to form nodules or to fix nitrogen, legumes have a high preceding crop effect in arable farming, which is highly valued by farmers. Figure 2.1.1-2 shows the nitrogen-fixing capacity of various grain legumes as a function of yield. However, most grain legumes have a low to very low self-tolerance, so that several years of cultivation break must be observed.

The production technique for growing grain legumes depends on the legume species and the location and can vary to a greater or lesser extent in different European countries depending on the machinery, the available varieties and the approved pesticides and fertilisers. Cultivation in Germany is therefore described as an example, whereby a distinction is made between conventional cultivation and organic farming in parts of the cultivation.

TABLE 2.1.1-1 Cultivation of grain legumes in the EU 2023 (in 1,000 ha)

Country	Faba beans	Grain peas	Sweet lupins	Soybeans	Total
Germany	61	118	25	45	249
France	80	152	5	158	395
Italy	4	15	1	311	331
Greece	50	19	12	1	82
Belgium	1	1			2
Ireland	16	1			17
Denmark	23	19			42
Spain	25	193	2	2	222
Austria	6	7	1	88	102
Finland	7	34			41
Sweden	20	27			47
Estonia	11	43			54
Latvia	32	36			68
Lithuania	80	71	4	1	156
Poland	37	33	157	18	245
Slovenia				3	3
Slovakia		15		50	65
Czech Republic		47	2	27	76
Hungary		12		58	70
Bulgaria		8		4	12
Romania	7	91		141	239
Croatia	3	1		71	75
Portugal	7		7		14
Slovenia					
Total EU-27	473	944	216	979	2,612

UFOP (2024), ACCORDING TO EU COMMISSION (2024)

2.1.2 *Cultivation*

2.1.2.1 Grain peas (*Pisum sativum L.*)

Of all grain legumes, the grain pea has the widest ecological spread. It produces the highest yields on humous, deep loamy soils with a neutral pH value of 6–7, but can also be grown on loamy sands and weathered soils if the water supply from rainfall is sufficient during growing time. Dry weather should prevail during ripening in July/August. Stony soil and surface characteristics are important when selecting a field. A quick and low-loss combine harvest is only possible on plane and stone-free areas.

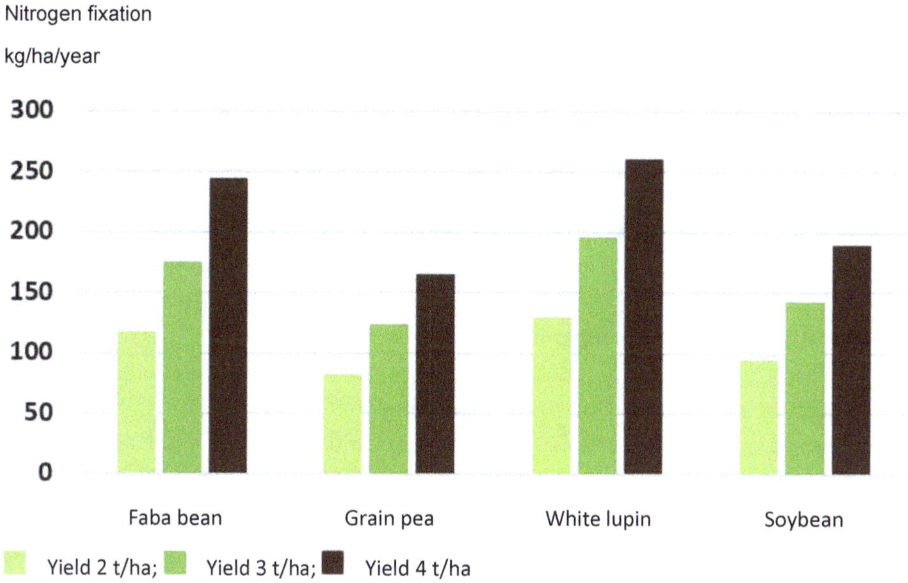

FIGURE 2.1.1-2 Nitrogen fixation of grain legumes as a function of yield (organic farming)
CALCULATED BY N-BUDGET CALCULATOR; © 2013 INSTITUTE OF LAND USE SYSTEMS, LEIBNIZ CENTRE FOR AGRICULTURAL LANDSCAPE RESEARCH, ZALF BY JOHANN BACHINGER, MORITZ RECKLING AND KARIN STEIN-BACHINGER (BACHINGER *ET AL.*, 2013)

FIGURE 2.1.2-1 Grain peas
PHOTO: © UFOP/D. SCHNEIDER

In the crop rotation, winter cereals are usually planted before and after the pea in order to break up cereal-rich crop rotations and to utilise the nitrogen left behind by the pea. Cultivation breaks of 6–9 years for white-flowering and 5–7 years for colour-flowering varieties should be observed between grain peas (Völkel and Vogt-Kaute, 2013).

In Germany, almost exclusively varieties of white-flowering, semi-leafless summer types are grown in conventional cultivation. White-flowering peas do not contain any tannins in their seeds. Initial practical trials are also being carried out with white-flowered semi-leafless winter peas, although the cultivation risk here is generally much greater depending on the location. In these varieties, only the stipules are developed. The leaflets are transformed into additional tendrils, so that the stands are more stable due to the better interlocking of the plants. Varieties of the normal type with stipules, developed leaflets and tendrils only at the end of the leaf stem, also colour-flowering, only play a role in organic farming in Germany. The selection of a variety with good vigour is crucial for successful cultivation.

Basic fertilising of peas is based on the principle of replacing nutrient removal from the field with the harvested crop (maintenance fertilising). As explained above, nitrogen fertilisation is not necessary for the grain pea and is not permitted in Germany under the Fertiliser Ordinance.

The spring type of peas can be sown from mid-March if no more late frosts are expected. The optimum sowing depth is 4–6 cm, the sowing density between 70–80 germinable seeds/m^2. Soil compaction and structural damage must be avoided at all costs during tillage and seedbed preparation, as the pea reacts very sensitively to this. Optimum soil condition therefore takes priority over the earliest possible sowing date. The row spacing in conventional cultivation can be freely selected from 12.5 (cereal spacing) cm to 30 cm for an intended machine hoe in organic farming. If available and possible, a fungicidal seed coating should be applied and the use of certified seed is recommended.

Winter peas are sown between mid-September and mid-November, depending on the location. Early sowing dates lead to more biomass, but reduce winter hardness. Sowing dates between the beginning and middle of October have therefore proven to be optimal.

Weed control is a critical point in the cultivation of peas. In Germany, only pre-emergence herbicides and graminicides are available for this purpose at the time of going to press. The time window for mechanical weed control – harrowing and hoeing is possible – closes at the time the plants become interlocked. As late growing weeds can be a problem in pea cultivation, it is important to ensure that the field is as weed-free as possible. In particular, it is important that there is little amount of free nitrogen from the previous crop, as this support weed growth.

In organic farming, peas are often cultivated as a mixed crop with cereals. Care must be taken to ensure that the selected crops or varieties are well synchronised in terms of ripening. The significant denser plant stand suppresses the weeds so that no weed control is necessary. Depending on the annual weather conditions, however, the proportion of the mixture partners in the crop can vary considerably. For targeted use in feeding, separation of the harvested crop is recommended, which induces additional costs.

Plant protection measures in conventional cultivation may require fungicide treatment and/or insecticide treatment. In this case, the pathogens must be monitored and a measure must be carried out if the damage threshold is exceeded. For further information, please refer to the "Guideline on integrated Plant Protection in the Cultivation of Faba bean, Grain pea, Soybean and Sweet lupins" from the Union zur Förderung von Oel- und Proteinpflanzen e. V. (Männel *et al.*, 2023).

Additional equipment on the cutting unit of the combine harvester (ear lifter, pick-up attachment, side knives) is advantageous for harvesting grain peas. Harvest time is reached when the plants are dry, the pods are light brown and the grains are hard. The optimum grain moisture for threshing is 15–17%. Grain peas can be stored at 12–14%, whereby care must be taken to ensure that spots with higher moisture levels do not form in the stack.

2.1.2.2 Faba beans (*Vicia faba L.*)

The faba bean has higher site requirements than grain peas. They need medium to heavy soils with a good water supply. Light soils and sites with regular pre-summer drought are not suitable for cultivation.

FIGURE 2.1.2-2 Faba beans
PHOTO: © UFOP/D. SCHNEIDER

Faba beans have no special requirements for the preceding crop. Like grain peas, they are usually grown after and before winter cereals. The cultivation break should be 4–5 years (Völkel and Vogt-Kaute, 2013).

Most faba bean varieties available on the seed market are colour-flowered, as these have a higher yield than white-flowered varieties. In contrast to white-flowered varieties, colour-flowered varieties contain tannins. In Germany, the spring type is almost exclusively cultivated, although farmers are also gaining initial practical experience with the cultivation of winter beans. The varieties available to date are more suited to a scenario with mild winters and pronounced early summer drought. On soils which are difficult to cultivate in spring, winter faba beans can be an alternative and higher-yielding than spring beans.

Basic fertilising of faba beans is based on the principle of replacing nutrient removal from the field with the harvested crop (maintenance fertilising). As explained above, nitrogen fertilisation is not necessary for field beans and is not permitted in Germany under the Fertiliser Ordinance.

Faba beans should be sown as early as possible in spring. Spring faba beans have a frost tolerance of −5 °C. They can therefore be sown from as early as February, but usually from the beginning of March. However, it is important that the soil is in "good condition", as faba beans react to unfavourable and excessively wet sowing conditions with poorer growth and lower yields. Faba beans require sowing depths of 6 to 10 cm. In practice, you should therefore try to sow them as deep as possible using the available sowing technology. Faba beans are sown in rows from 12.5 cm (cereal spacing) up to 45 cm row spacing. In Germany, higher sowing rates of 40–45 germinable seeds/m^2 are recommended, as such crops have better yield stability. The modern varieties have good to very good stability, so there is no need to worry about an increased tendency to lodging. If available and possible, fungicidal seed coating should be applied and the use of certified seed is recommended.

Depending on the location, winter beans are sown in Germany from September 20 to the end of October with 25–30 germinable seeds/m^2. It is important that the plants have approx. 6 weeks of growing time until the end of vegetation. As with the winter pea, the winter hardness of the winter faba bean decreases as the biomass increases.

At the time this book went to press, only pre-emergence herbicides and graminicides were still available for weed control in Germany. Faba beans are very well suited to mechanical weed control using harrows and hoes. As late growing weeds can also be a problem in faba bean cultivation, it is important to ensure that the field is as weed-free as possible. In particular, it is important that there is little free nitrogen amount from the previous crop, as this encourages weed growth.

Plant protection measures in conventional cultivation may require fungicide treatment and/or insecticide treatment. In this case, the pests must be monitored and measures must be taken if the damage threshold is exceeded. For further information, please refer to the "Guideline on integrated Plant Protection in the

Cultivation of Faba bean, Grain pea, Soybean and Sweet lupins" from the Union zur Förderung von Oel- und Proteinpflanzen e. V. (Männel *et al.*, 2023).

Modern faba bean varieties ripen in Germany in August or by mid-September at the latest. They are ready for threshing when the plants, including the pods, are black. The proportion of pods that are still green should not exceed 10%. To facilitate collection and processing, the cutting unit of the combine harvester is set as high as possible so that it starts just below the lowest pods. A wide cutting unit and a high driving speed support cutting and intake and minimise grain losses. Faba beans with a moisture content of 12–14% are suitable for storage, whereby care must be taken to ensure that spots with a higher moisture content do not form in the stack.

2.1.2.3 Sweet lupins (*Lupinus angustifolius L.* and *Lupinus albus L.*)

In the case of sweet lupins, a distinction must be made between blue or narrow-leaved, white and yellow lupins. While low-bitter or sweet varieties of white and yellow lupins have been available for a long time, plant breeders only achieved this breakthrough with blue or narrow-leaved lupins in 1997. The different types of lupin have different requirements in terms of soil and climate. The yellow lupin is very similar to rye in its requirements, the blue or narrow-leaved lupin to barley and the white lupin to wheat. Lime sensitivity in particular is a criterion for site selection. The pH value of the field should be between 5 and 6. White lupin is much more tolerant here and accept higher pH value. The yield potential increases from the yellow to the blue or narrow-leaved to the white lupin. However, the yellow lupin has the highest protein content in the grain and is also suitable for cultivation on very sandy and dry sites. In Germany, however, only blue or narrow-leaved and white sweet lupins are currently cultivated due to the fungal disease anthracnose. Blue or narrow-leaved sweet lupins are generally less susceptible to anthracnose, while new anthracnose-tolerant varieties of white lupin were approved in 2019 (BSA, 2023). In other EU member states, yellow sweet lupin also plays a role in practical cultivation, e. g. an anthracnose-resistant variety of yellow sweet lupin is known from Poland and research is already underway to transfer this resistance into modern breeding material from German lupin breeders (Haase *et al.*, 2023).

In the crop rotation, lupins are usually preceded and followed by winter cereals in order to break up cereal-rich crop rotations and to utilise the nitrogen left behind by the lupins. Cultivation breaks of 4–5 years should be observed between lupins (Völkel and Vogt-Kaute, 2013). The cultivation areas of bitter and sweet lupins must be strictly separated, as lupins are hard-shelled and can continue to grow for a long time after cultivation. The bitter substance content in bitter lupins is so high that very high bitter substance levels can be induced in sweet lupin plots with just a few volunteer plants per hectare. Farmers growing sweet lupins should also ensure that their intercrop mixtures do not contain any lupins, as these are usually bitter lupins.

FIGURE 2.1.2-3 Sweet lupins
PHOTO: © UFOP/M. RUNGE

The basic fertilising of lupins is based on the principle of replacing the removal of nutrients from the field with the harvested crop (maintenance fertilising). As explained above, nitrogen fertilisation is not necessary for lupins and is not permitted in Germany under the Fertiliser Ordinance.

Lupins should be sown in spring from the middle/end of March in well-warmed soil with the aim of rapid emergence. A slightly delayed sowing is acceptable. The sowing depth is 2 to 3 cm. The seed quantity for blue or narrow-leaved sweet lupin varies between 90–100 for the branching type and between 90–120 germinable seeds/m^2 for the terminal type. For the white lupin, the sowing rate is 70–80 for the branching type and between 80–90 germinable seeds/m^2 for the terminal type. The row spacing is 12.5 cm (grain spacing). In the case of mechanical weed control, a higher spacing can also be selected depending on the machine used. When using a hoe, double the crop spacing has proved effective. With very wide rows, the crops close much later and the risk of late growing weeds increases. These should therefore be avoided wherever possible, as lupins are very low resistant to competition from weeds. If available and possible, a fungicidal coating should be applied. In Germany, the reproduction of white and blue or narrow-leaved lupins is prohibited, so that certified seed must be used.

At the time this book went to press, only pre-emergence herbicides and graminicides were still available for weed control in Germany. Sweet lupins are very well suited to mechanical weed control with a harrow. As late growing weeds can also be a problem in lupin cultivation, it is important to ensure that the field is as weed-free as possible. In particular, it is important that there is little amount of free nitrogen from the previous crop, as this support weed growth.

Insecticide treatment may be necessary as a plant protection measure in conventional cultivation – the use of fungicides against anthracnose is usually not economical. The pests must be monitored and measures must be taken if the damage threshold is exceeded. For further information, please refer to the "Guideline on integrated Plant Protection in the Cultivation of Faba bean, Grain pea, Soybean and Sweet lupins" from the Union zur Förderung von Oel- und Proteinpflanzen e. V. (Männel *et al.*, 2023).

Lupins are harvested when 95% of the pods and stalks are brown. In warm and dry weather, the harvest date for blue or narrow-leaved sweet lupins must have highest priority, as the pods can burst open very easily. In order to avoid harvest losses, the lupin crop must not be allowed to overripe under any circumstances, so that the entire combine harvester capacity of the farm should be utilised. Threshing at night and in the morning has proven to be very effective. Lupins with a moisture content of 12–14% are suitable for storage, whereby care must be taken to ensure that there are no spots with higher moisture content in the stack.

2.1.2.4 Soybeans (*Glycine max L.*)

Compared to the grain legumes described above, the soybean is still a relatively young crop in many EU member states, especially in Central Europe. The soybean prefers easily warmed soils, for example warm grain corn cultivation sites. As the water requirement is particularly high from flowering and during grain development (July/August), sufficient and continuous water availability is very important for assessing the suitability of the site – sites with field bindweed (*Convolvulus arvensis L.*) problems are ruled out. Fields for soybean cultivation must be free of stones to avoid problems during harvest.

The soybean has no special requirements for the preceding crop. Like other grain legumes, they are usually grown after and before winter cereals. When selecting the following crop, the maturity date of the soybean must be taken into account, which can be in September or later depending on the region and year. For reasons of arable hygiene, the cultivation break should be 2–3 years despite the good self-tolerance, even if there are currently no known crop rotation diseases in soybean in Europe (Völkel and Vogt-Kaute, 2013).

The basic fertilising of soybeans is based on the principle of replacing the removal of nutrients from the field with the harvested crop (maintenance fertilising). As explained above, nitrogen fertilisation is not necessary for soy and is not permitted in Germany under the Fertiliser Ordinance.

A timely sowing date is important for successful soy cultivation, as soon as the soil has warmed up to at least 10 °C by mid-April. In conventional cultivation, soy is often sown at a grain spacing (12.5 cm) or half a maize spacing (37.5 cm). Especially

FIGURE 2.1.2-4 Soybeans
PHOTO: © UFOP/D. SCHNEIDER

with early maturing varieties with low branching too wide row spacing costs yield. The sowing rate is 55 germinable seeds/m^2. The sowing depth is 3–4 cm on heavy soils and 4–6 cm on light soils. Under no circumstances should the soybean be sown too deep, as this will result in insufficient emergence.

The basis for clean soy fields is thorough seedbed preparation. The most important leading weed in soy is the white goosefoot (*Chenopodium album L.*), as it can lead to massive competition and thus to high yield losses. It is immediately followed by black nightshade (*Solanum nigrum L.*). However, the soy plant can react with reduced yields if herbicide application rates are not adapted to the conditions. At the time this book went to press, both pre-emergence and post-emergence herbicides and graminicides were available in Germany, which can be combined into a spray sequence if necessary. Post-emergence weed control alone is not possible. Weed control in soybeans can also be achieved mechanically by hoeing and harrowing. However, harrowing alone is usually not sufficient in organic farming with cereal spacing. In soy cultivation, care must be taken to ensure that the field is as weed-free as possible. In particular, it is important that there is little amount of free nitrogen from the previous crop, as this support weed growth.

With the exception of herbicide treatment, hardly any plant protection measures are necessary for soy cultivation with the current level of cultivation in Germany, as this crop is still considered to be very healthy. In individual years, it may be necessary to control the painted lady (*Vanessa cardui L.*). For further information, please refer to the "Guideline on integrated Plant Protection in the Cultivation of Faba bean, Grain pea, Soybean and Sweet lupins" from the Union zur Förderung von Oel- und Proteinpflanzen e. V. (Männel *et al.*, 2023).

The soybean harvest date is reached when the stalks turn reddish-brown, the pods rustle and the seeds can no longer be pressed in. Depending on the location, annual weather conditions and chosen maturity group, this is usually the case between mid-September and early October. When threshing soy, it is important to note that the pods are very low and that the grains break easily. This must be taken into account when equipping the combine harvester (flex cutting unit if possible) and adjusting it. The bursting of pods before harvesting is generally no longer a problem with modern varieties. Corn, peas, faba beans and similar crops are difficult to clean out of soybeans after harvest. Soybeans with a moisture content of 12–14% are suitable for storage, although care must be taken to ensure that there are no spots of higher moisture in the stack.

References

Bachinger J., Reckling M., Stein-Bachinger K., 2013: N-Fixierleistung von Körnerleguminosen, Berechnet mit dem N-Budget Calculator. https://www.nutrinet.agrarpraxisforschung.de/naehrstoffmanagement/n-versorgung-verbessern/anbau-von-koernerleguminosen#:~:text=Die%20N%2DFixierleistung%20von%20K%C3%B6rnerleguminosen%20ist%20in%20der%20Regel%20deutlich,30%20und%2040%20dt%2Fha. Retrieved July 10, 2024.

BSA, 2023: Beschreibende Sortenliste Getreide, Mais, Öl- und Faserpflanzen, Leguminosen, Rüben, Zwischenfrüchte 2023. https://www.bundessortenamt.de/bsa/media/Files/BSL/bsl_getreide_2023.pdf. Retrieved July 4, 2024.

DESTATIS, 2010–2023: Fachserie / 3 / 3 / 2 / 1. 2010,09 – Feldfrüchte. 2011,16 – Feldfrüchte. 2012, 16 – Feldfrüchte. 2013,16 – Feldfrüchte. 2014, 16 – Feldfrüchte. 2015,16 – Feldfrüchte. 2016, 16 – Feldfrüchte. 2017,16 – Feldfrüchte. 2018,09 – Feldfrüchte. 2019,09 Feldfrüchte. 2020,09 – Feldfrüchte. 2021,09 – Feldfrüchte. 2022,09 – Feldfrüchte. Ackerland nach Hauptfruchtgruppen und Fruchtarten. Wiesbaden. https://www.statistischebibliothek.de/mir/receive/DESerie_mods_00000335 and https://www.destatis.de/DE/Themen/Branchen-Unternehmen/Landwirtschaft-Forstwirtschaft-Fischerei/Feldfruechte-Gruenland/Tabellen/ackerland-hauptnutzungsarten-kulturarten.html. Retrieved July 12, 2024.

DESTATIS, 2024: https://www.destatis.de/DE/Themen/Branchen-Unternehmen/Landwirtschaft-Forstwirtschaft-Fischerei/Feldfruechte-Gruenland/Tabellen/ackerland-hauptnutzungsarten-kulturarten.html. Retrieved July 2, 2024.

Haase F., Krebes C., Lohwasser U., Dieterich R., Eickmeyer F., Fetzer A., Ruge-Wehling B., 2023: Increasing the cultivability of Lupinus luteus: anthracnose resistance and yield potential. In: International Lupin Conference XIV "Breeding, cultivation and use of lupins for a sustainable agriculture – recent developments", 19.–23. June 2023 in Rostock, Germany; Book of Abstracts, p. 34.

Männel M., Schäfer B.C., Haberlah-Korr V., 2023: Leitlinie des integrierten Pflanzenschutzes beim Anbau von Ackerbohne, Körnererbse, Sojabohne und Süßlupinen. https://www.ufop.de/files/9616/8447/2921/UFOP_2096_Leitlinie_Kornerleguminosen_DINA4_160523.pdf. Retrieved July 3, 2024.

UFOP, 2024: Geschäftsbericht 2023/24. https://www.ufop.de (unpublished at the time of going to press).

Van Loon M.P., Alimagham S., Pronk A., Fodor N., Ion V., Kryvoshein O., Kryvobok O., Marrou H., Mihail R., Mínguez M.I., Pulina A., Reckling M., Rittler L., Roggero P.P., Stoddard F.L., Topp C.F.E., van der Wel J., Watson C., van Ittersum M.K., 2023: Grain legume production in Europe for food, feed and meat-substitution. Global food security, 39, 100723. doi: 10.1016/j.gfs.2023.100723.

Völkel G., Vogt-Kaute W., 2013: Körnerleguminosen anbauen und verwerten. KTBL-Heft 100, pp. 7–9 und 24–25.

2.2 Grain legumes in aquaculture

Monika Weiß, Vanessa Fuchs and Sinem Zeytin

The quality of raw materials for inclusion in aquafeed is of extraordinary importance, because most commercially important species are produced in intensive or semi-intensive culture, where nutrition is exclusively reliant on accurately formulated aquafeeds, especially designed to meet the requirements of the target species at each applicable life-stage. These feeds consist of a variety of ingredients for which consistent quality and nutrient composition are indispensable for reliable feed production. Fishmeal remains one of the most important resources for feed production, which has an advantageous amino acid profile, is highly digestible and exhibits extremely high palatability. The demand, availability and price led to rapid replacement of fishmeal in aquafeeds during the early stages of aquaculture's rapid expansion since 1990. Legumes proved to be an adequate source for fishmeal replacement and primarily the inclusion of soybeans made a profound reduction of fishmeal in aquafeeds possible. Developments in recent years have shown that it is advisable to diversify towards a repertoire of regionally available resources in order to strengthen the resilience of the aquafeed sector. Researchers worldwide dedicated applied research to this task and available grain legumes have been tested for their applicability in aquafeeds. As a result, it became increasingly evident that despite legumes' generally successful inclusion in aquafeeds, anti-nutritional factors (ANF) can limit the applicability or demand further processing of legumes

(Drew *et al.*, 2007) and their inclusion generally entails an addition of single amino acids (mainly sulfur-containing amino acids, which are limited in legumes) to complete the required amino acid profiles.

Another relevant topic that to date receives limited attention, but is of uttermost importance in fish feed, is the manufacturing processes. Legumes processing, binding and extrusion behavior is dependent on the legume and the product applied (e.g. protein concentrate/isolate). The feed production processes have to be adjusted depending on the resource to achieve the required physical properties of the feed pellet (Kraugerud *et al.*, 2011). This not only concerns durability and pellet water stability, but also strongly affects the digestibility of a feed, especially concerning the reduction of antinutritional factors. The whole field of physical – physiological interactions of ingredient dependent feed pellet quality, including faeces properties, is receiving an increasing amount of attention, and will improve feed quality in terms of the reduction of feed losses immensely during the next years.

2.2.1 *General remarks*

A severe obstacle when interpreting the results of the various existing studies on the application of alternative resources is the divergence in the experimental approach. Some authors replace FM, some soybean products, some protein sources in general, some carbohydrate sources. In many of the older studies one source is just replaced with the other 1:1, later studies formulate the experimental feeds as isonitrogenous and/or isocaloric but without balancing essential amino acids. Most successful in the application of alternative resources are studies that precisely formulate the feeds to the species and life stage demands and balance missing amino acids (mainly sulfur-containing amino acids like methionine) with artificial products. Currently, the inclusion of these products excludes the feed and resulting aquaculture products from an eco-certification, which is obviously an issue that has to be solved, because this restriction finally hinders the commercial development of sustainable, eco-certified feedstuff.

The outcome of a study is often dependent on the formulation of the control feed, upon which the whole study is based. In many studies, diets are formulated with a surplus of protein mainly from fishmeal, which often cloaks the effects of replacement with a legume or another alternative resource tested, since test diets might still contain a higher percentage of fishmeal than a commercial diet would. A related problem that occurs in studies where the amino acid profile is not balanced, is the application of fishmeal of different qualities: If a high-quality fishmeal is replaced with a legume, the results for the diet with legumes can easily be worse than for the control diet. If the fishmeal is of lower quality, the results can very easily be the exact opposite. As a recommendation for future feeding experiments: i) diets should be formulated to meet the minimum required protein level to clearly reveal deficiencies. ii) the essential nutrients, not only amino acids, but also fatty acid composition, vitamins and minerals should be balanced with special care, since it is of uttermost importance for animals' health and growth. In this sense, the list of the 10 "essential" amino acids for fish have to be expanded, since e.g. Taurin is considered non-essential but cannot be synthesized sufficiently, especially by clearly piscivore species.

2.2.2 *General recommendations*

Ingredients should be analyzed for their nutritional profile before a feed is formulated, as the type of cultivation and the cultivation region and cultivar show differences in the nutrient profile. It is recommendable to use a combination of different (protein) resources (e.g. fishmeal, meat meal, blood meal, grain peas, sweet lupins), which leads to balanced the amino acid profile on the one hand and is significantly more cost effective than other diets (Allan *et al.*, 2000a; Allan *et al.*, 2000b; Stone *et al.*, 2000; Booth *et al.*, 2001; Stone *et al.*, 2003).

References

Allan G.L., Parkinson S., Booth M.A., Stone D.A.J., Rowland S.J., Frances J., Warner-Smith R., 2000a: Replacement of fish meal in diets for Australian silver perch, *Bidyanus bidyanus*: I. Digestibility of alternative ingredients. Aquaculture 186(3–4): 293–310.

Allan G.L., Rowland S.J., Mifsud C., Glendenning D., Stone D.A.J., Ford A., 2000b: Replacement of fish meal in diets for Australian silver perch, *Bidyanus bidyanus*: V. Least-cost formulation of practical diets. Aquaculture 186(3): 327–340.

Booth M.A., Allan G.L., Frances J., Parkinson S., 2001: Replacement of fish meal in diets for Australian silver perch, *Bidyanus bidyanus*: IV. Effects of dehulling and protein concentration on digestibility of grain legumes. Aquaculture 196(1): 67–85.

Drew M.D., Borgeson T.L., Thiessen D.L., 2007: A review of processing of feed ingredients to enhance diet digestibility in finfish. Animal feed science and technology 138(2): 118–136.

Kraugerud O.F., Jørgensen H.Y., Svihus B., 2011: Physical properties of extruded fish feed with inclusion of different plant (legumes, oilseeds, or cereals) meals. Animal feed science and technology 163(2): 244–254.

Stone D.A.J., Allan G.L., Parkinson S., Frances J., 2003: Replacement of fishmeal in diets for Australian silver perch *Bidyanus bidyanus* (mitchell). II. Effects of cooking on digestibility of a practical diet containing different starch products. Aquaculture research 34(3): 195–204.

Stone D.A.J., Allan G.L., Parkinson S., Rowland S.J., 2000: Replacement of fish meal in diets for Australian silver perch, *Bidyanus bidyanus*: III. Digestibility and growth using meat meal products. Aquaculture 186(3): 311–326.

2.3 Antinutritive factors (ANF) in grain legumes

Mechthild Freitag

All legumes contain organic constituents that may limit their acceptance in diets or may affect animal health. Therefore, they are called "antinutritive factors" (ANF). These substances are not essential for plant growth and survival. They are formed as secondary metabolites to protect plants from herbivores, to increase their resistance to microbial infections, to regulate plant growth, or to provide colour or aroma. Their concentration varies according to the species and variety of legume, but also according to the weather and soil conditions under which they grow. Unfortunately, there is no visible sign of higher ANF levels on the seeds. Their effects become apparent when feeding compounds with these legumes. Restricted feed intake is often the first sign of high ANF levels. However, it has to be kept in mind that reduced feed intake can also be caused by poor hygienic conditions in the feed, e.g. moulds, or some mycotoxins.

ANF belong to different chemical subgroups.

2.3.1 *Alkaloids*

Alkaloids are basically water soluble and organic nitrogen containing substances. Most of them are alkaline. They are characterised by a bitter taste and some toxic potential. This bitter taste may reduce feed intake and thus limit the rate at which a legume seed can be included in animal diets.

Quinolizidine alkaloids are predominantly found in lupins. Most problematic are lupinin, sparteine and lupanine and its isomers. Further major quinolizidine alkaloids in lupins are angustifoline, albine and 3-hydroxylupanine. Chemically, quinolizidine alkaloids are based on a bicyclic Q ring system.

High levels (alkaloid concentration 0.1 to 2%) – as observed in bitter lupins – cause toxicity through neurological effects as antagonists to acetylcholine receptors and by inhibiting Na^+ and K^+ channels and thus may lead to loss of motor co-ordination and muscular control. Besides disturbances of the central nervous system, further toxic effects of alkaloids result in negative effects on digestive processes, reproduction and the immune system. Antipalatability effects of lupin

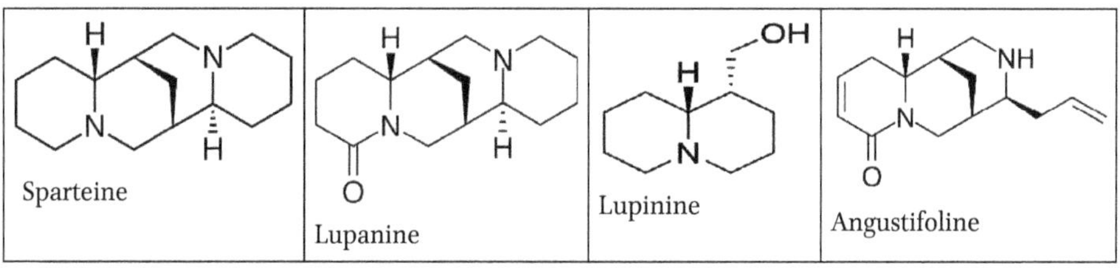

FIGURE 2.3.1-1 Chemical structures of quinozidine alkaloids

alkaloids are also observed and may be mediated in part by neurological effects. Reduced feed intake is usually the obvious effect of higher alkaloid levels.

Quinolizidine alkaloids can be reduced or even eliminated by heating, sprouting and fermentation. Beneficial for plants is their nitrogen content as a reserve during sprouting. Moreover, their bitter taste scares herbivores off.

Concentrations are high in bitter lupin cultivars, but are generally low in sweet lupins as well as in soya, faba beans and peas. However, if reduced feed intake is observed when feeding diets with legumes these ingredients should be analysed for their alkaloid content.

2.3.2 *Chelating agent: phytic acid*

Phytic acid, inositol hexakisphosphate (IP6), is the only chelating agent of interest in legumes. Phytic acid and its salts represent the majority of phosphorus (P) in legume seeds. Without further treatment, this P is not digested and therefore not available to the animals.

Phytate accumulates in the seeds of all grain legumes during the ripening period (table and figure 2.3.2-1). The level in seeds varies depending on the legume species, variety, and growing conditions like weather and soil. Analyses revealed that phytic acid may constitute up to 78% of the total P content of the legume species, with highest levels in soybean seeds (Chitra *et al.*, 1995).

To a minor extent, phytic acid also binds to Ca^{++}, Mg^{++}, Zn^{++} and Fe^{++}. It also complexes with proteins and thus reduces their digestibility. Moreover, phytates impaire enzyme activity with negative effects for digestive enzymes like amylase,

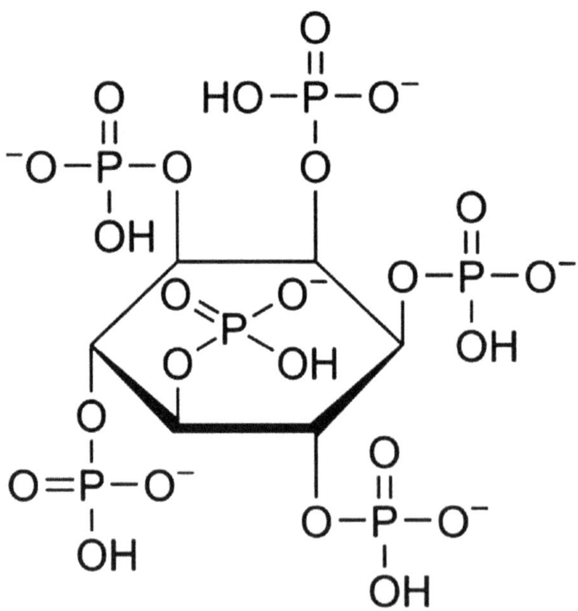

FIGURE 2.3.2-1 Chemical structure of phytate

TABLE 2.3.2-1 Phytic acid in selected grain legumes

Legume species	Phytic acid (mg/g)	Authors
Glycine max L.	12.7	Egounlety and Aworth, 2003, modified
Vicia faba L.	1.7–21.25	Meng *et al.*, 2021
Pisum sativum L.	7.3	Elango *et al.*, 2022
Lupinus albus L.	8.2	Elkowitz and Sosulski, 1982

pepsin and trypsin and thus reduce overall nutrient digestibility. In ruminants, these complexes are hydrolysed during rumen fermentation and minerals and proteins are made available for resorption. Monogastric species, however, have a limited ability to hydrolyse phytate. Therefore, the phytic acid level limits nutrient availability in monogastric animals. Depending on the phytic acid content, protein and energy digestibility are reduced and P emissions enhanced.

The elimination of negative effects is possible by adding phytase enzymes to monogastric diets. This is common practise in pig and poultry feed in many countries. The phytate content in seeds can also be reduced by soaking or by fermentation of legumes, especially if they have sprouted before.

2.3.3 *Glucosides and glycosides: vicine, convicine and saponins*

Chemically, glucosides are a group of organic components in which an alcohol or phenol (R-OH) is bound to glucose via a glycosidic bond. Glucosides are a subgroup of the glycosides. There are four mayor glucosides in legumes: vicine, convicine, saponins and α-galactosides.

The most important glucosides in legume seeds are vicine and convicine (V/C) in faba beans. They are thermostable glucosidic aminopyrimidine derivatives. They accumulate at maturity in the cotyledons of faba beans (*Vicia faba L.*) and vetches (*Vicia*). The content of vicine and convicine varies greatly between plant varieties. In conventional cultivars carrying the vc + allele, the content ranges from 6 to 14 g/kg DM, whereas homozygous genotypes negative for the vc allele have a 10 to 20-fold reduced content (Crépon *et al.*, 2010). Both glycosides can be hydrolysed in the digestive tract to divicine and isouramil. These substances can oxidise DNA, proteins and lipids and thus cause disorder of the fat metabolism. Vicine and convicine act by reducing glutathione and glucose-6-phosphate dehydrogenase activity, which may result in haemolytic anaemia due to biochemical abnormalities of blood cells.

Both, but particularly vicine, have strong effects on laying hens, reducing feed consumption, weight gain, egg weight and fertility and hatchability of eggs. A long exposure (140 days) to hens with 20% inclusion of field beans with high vicine and convicine content (1.17 g/kg DM) resulted in significant reduced feed intake, less

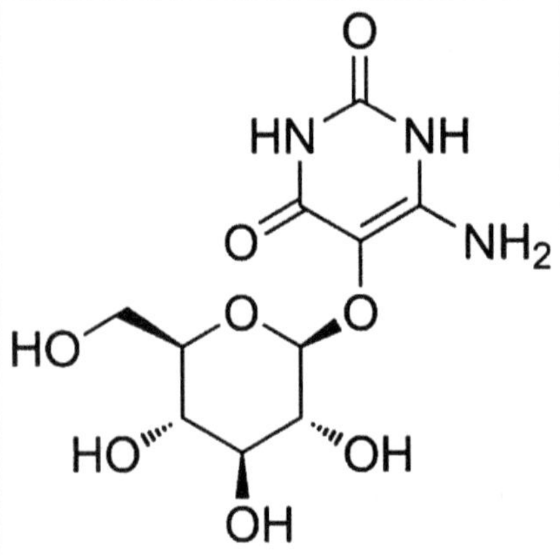

Convicine

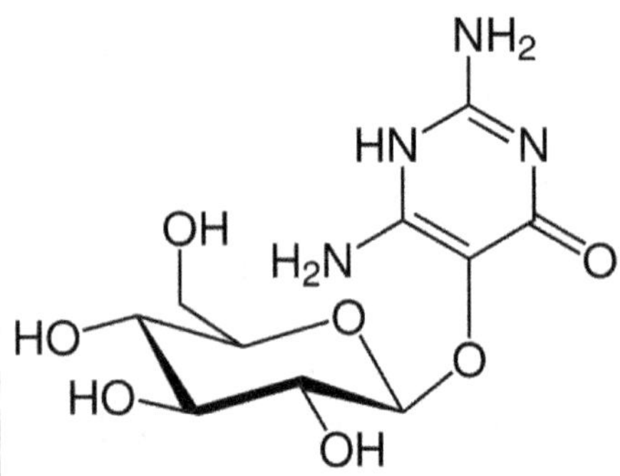

Vicine

FIGURE 2.3.3-1 Chemical structures of vicine and convicine

TABLE 2.3.3-1 Saponin content of some legume species (g/kg DM)

Legume species	Saponin content
Soybean meal	49.4
Faba beans	
Colour-flowered	25.5–39.3
White-flowered	16–18
Sweet lupins	
L. albus	< 0.012
L. luteus	0.055
L. angustifolius	0.379–0.740

JEZIERNY *ET AL.*, 2010

mean egg weight and reduced egg mass (Lessire *et al.*, 2017). Mean egg weight and egg mass were already significantly reduced at medium vicine and convicine content (0.45 mg/kg DM; $p < 0.003$). The number of eggs per hen was not affected. It appears that growing chickens are less affected than laying hens. Their feed intake was not impaired (Muduuli *et al.*, 1982). In laying hens, even 10 g vicine/kg diet may limit feed intake and reduce egg weight, fertility and hatchability of eggs, packed cell volume and erythrocyte haemoglobin levels and, in addition, may lead to increased liver weights, liver glutathione levels, liver and plasma lipid levels, plasma lipid peroxide levels and erythrocyte haemolysis in vitro (Jezierny *et al.*, 2010).

Saponins are characterised by an astringent taste, and their antinutritive effect is likely due to an increased permeability of the small intestinal mucosa cells which, in turn, may lead to an inhibition of active nutrient transport across the intestinal wall. Moreover, saponins also have a bitter taste that may limit feed intake in monogastric animals. They are mainly present in soybeans, up to 49.4 g/kg and in colour-flowered faba beans (table 2.3.3-1). Peas and lupins contain only negligible amounts (Jezierny *et al.*, 2010). In lupins, the saponin content is positively correlated with their alkaloid content. However, as their content is generally very low, they will not restrict feed intake.

2.3.4 *Isoflavones*

Several different isoflavones can be found in legumes. The main components – found predominantly in soybeans (*Glycine max L.*), red and white clover (*Trifolium pratense L., Trifoiums repens L.*) and lucerne (*Medicago sativa L.*) – are genistein,

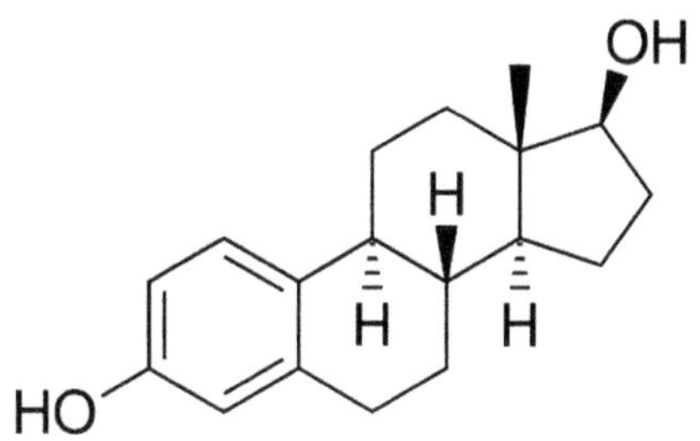

17β-Estradiol

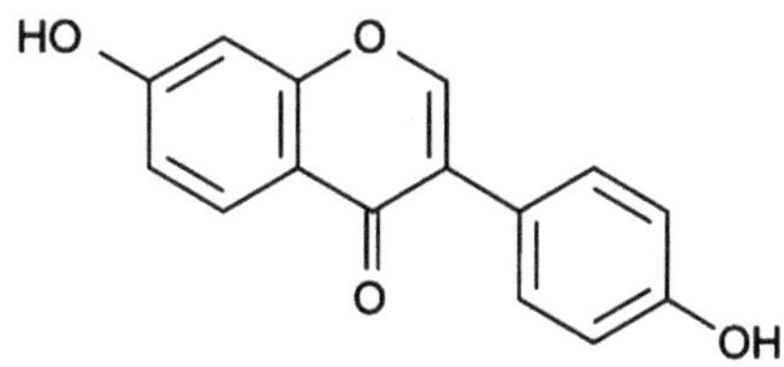

Daidzein (7,4-dihydroxyisoflavone)
Slashme, CC BY-SA 3.0 via Wikimedia Commons

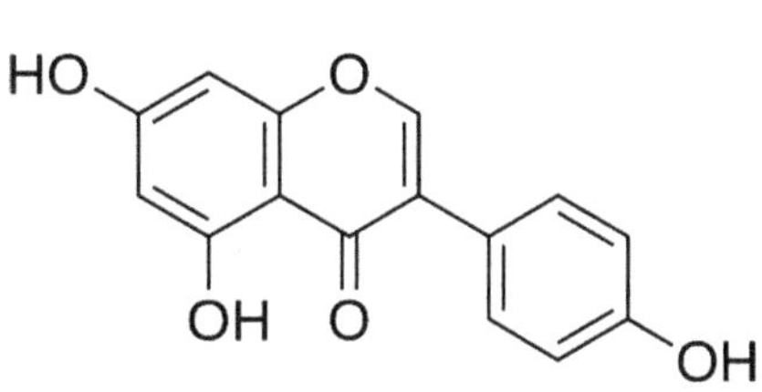

Genistein, (7,4'-dihydroxy-6-methoxyisoflavone)

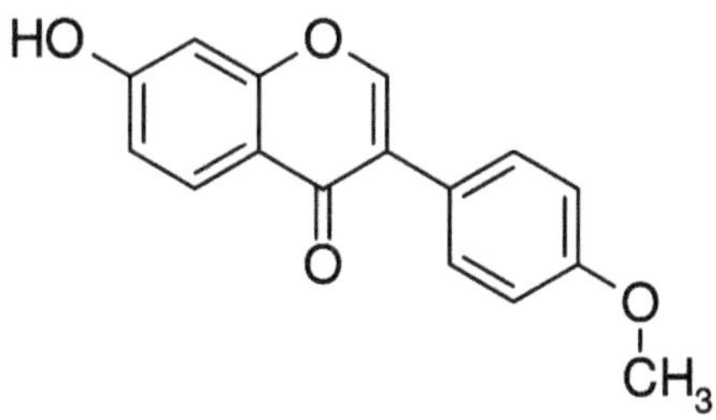

Formononetin (7-hydroxy-4′-methoxyisoflavone)

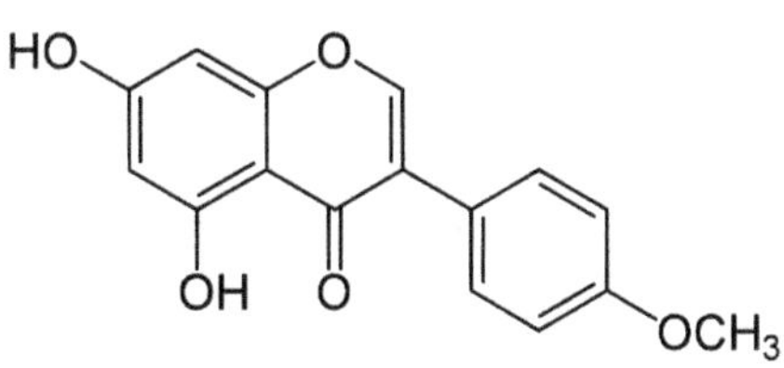

Biochanin A (5,7-dihydroxy-4-methoxyisoflavone)
Yikrazuul, Public domain, via Wikimedia Commons

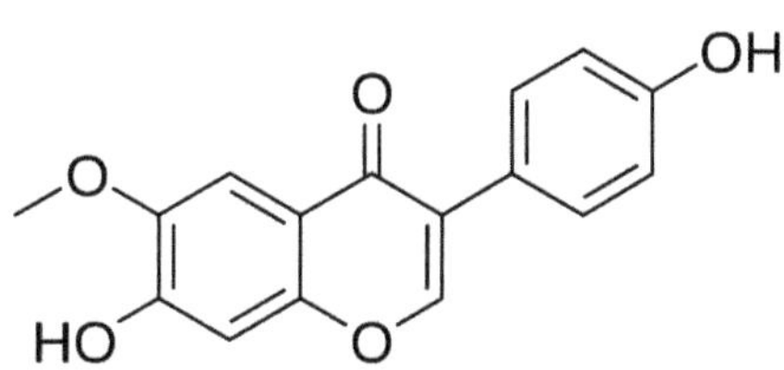

Glycitein (7,4-dihydroxy-6-methoxyisoflavone)

FIGURE 2.3.4-1 Chemical structures of isoflavones

TABLE 2.3.4-1 Isoflavone content of some legume species

Species	Content (mg/g dry matter)
Soybeans (*Glycine max L.*)	1.2–4.2
Red clover (*Trifolium pratense L.*)	10–25
White clover (*Trifolium repens L.*)	0.5–0.6
Lucerne (*Medicago sativa L.*)	0.05–0.3

KRÍŽOVÁ *ET AL.* (2019)

daidzein and glycitein in soybeans and formononetin, biochanin A, daidzein and genistein in red clover.

Isoflavones are ranked among the most estrogenic plant molecules. Phytoestrogens are naturally occurring nonsteroidal phenolic components that, due to their molecular structure and size, resemble vertebrate steroids and can replace oestrogen at receptors. In plants, they serve as defence mechanisms against several, primarily fungal, pathogens. The concentration of phytoestrogens in a plant depends mainly on the plant variety, the growth stage, plant part and growing conditions. Isoflavone concentration raises sharply during stress, e.g. drought, pathogen attack, or plant diseases and is influenced by environmental conditions like weather, soil quality or harvest period. It may also be influenced by the post-harvest processing and conservation method.

In animals, estrogenic flavonoids cause reproductive disorders in females, predominantly in sheep, but also in cows, which may be permanent after a flavonoid exposure for a long time. Effects are early embryonic mortality, miscarriage and stillbirth in sheep and in young cattle. Older cows were not affected.

On the other hand, positive effects from flavonoid uptake could also be detected. In dairy cows, higher feed intake and milk production was observed as well as improved weight gain in lambs and improved laying performance in hens (Křížová *et al.*, 2019). More research into these effects is needed.

2.3.5 *Oligosaccharides*

Legume seeds contain α-galactosides, such as sucrose, raffinose, stachyose and verbascose. They belong to the non-digestible oligosaccharides because the vertebrate digestive tract does not produce α-galactosidase. Thus, they reduce overall digestibly of legume nutrients. Moreover, they cause flatulence in humans and animals.

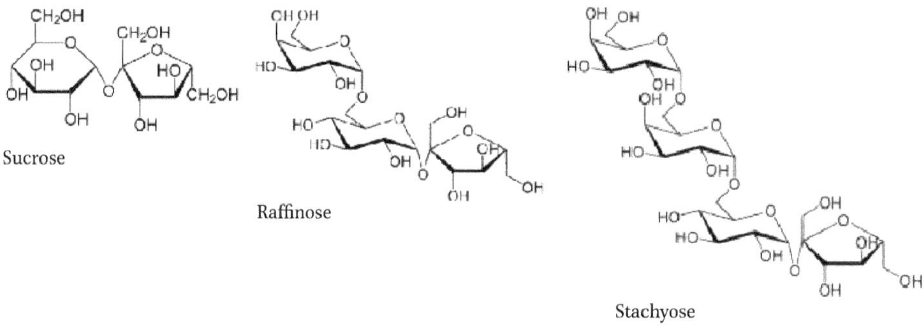

FIGURE 2.3.5-1 Chemical structure of some oligosaccharides

TABLE 2.3.5-1 α-Galactosides in legume seeds (g/100 g)

Species	Total α-galactosides	Sucrose	Raffinose	Stachyose	Verbascose
Soybean seed	6.0–8.0	3.6–6.1	0.6–1.1	2.2–4.4	0.2
Soybean meal	0.4–12.5	0.5–7.7	0.1–3.6	2.9–4.1	–
Grain peas	2.3–9.6	1.2–2.8	0.3–1.2	1.1–3.8	0.0–4.6
Faba beans	2.2–3.7	1.4–2.0	0.0–0.3	0.7–1.1	1.2–2.3
Sweet lupins					
L. albus	5.9–8.7	2.6–2.9	0.8–0.9	4.1–6.6	0.5–1.1
L. luteus	8.6–10.2	1.7–3.4	1.2–1.5	4.9–5.2	2.0–4.1

ELANGO *ET AL.* (2022)

Flatulence is an important factor that deters consumption and utilization of legumes in human and animal diets.

α-Galactosides are rapidly fermented by microorganisms in the digestive tract, foremost in the large intestine. The fermentation is accompanied by hydrogen, methane, and CO_2 production, thus leading to flatulence.

Raffinose family oligosaccharides are soluble carbohydrates. Their concentrations in seeds are related to the environment, genotype, and harvest time. In plants, they are a form of carbohydrate storage and transport. Thus they play an important role in abiotic stresses, such as high salinity and drought. They have been reported to impart also tolerance to heat, cold, and disease resistance besides regulating seed germination, vigour, and longevity (Elango *et al.*, 2022). Raffinose oligosaccharides can be found in a variety of legume seeds (table 2.3.5-1).

The reduction of α-galactosides in feed ingredients is possible through a variety of applications, such as soaking, germination, decortication, fermentation, heating and addition of the enzyme α-galactosidase. Moreover, breeding varieties with low content of α-galactosides has been successful, thus creating varieties with a higher nutritional value by enhanced digestibility. These varieties are recommended for feeding purposes.

On the other hand, some indigestible oligosaccharides have recently been reported to have a beneficial effect on the gut microflora. They promote the growth of beneficial bacteria such as bifidobacteria and lactobacilli and reduce the harmful bacteria present in the colon (Elango *et al.*, 2022). In future, special legume cultivars will probably be included into the diet to serve as prebiotics.

2.3.6 *Tannins*

Tannins are condensed polyphenols with a sufficiently large number of phenolic hydroxyl groups to form cross-linkages with proteins and other macro-molecules. Condensed tannins are a subset of this group. They are the most abundant type of tannins in economically important plants, occurring mainly in the coats of leguminous seeds and in legume forages. They form complexes with dietary nutrients and make them resistant to the action of digestive enzymes. They can also inactivate digestive enzymes themselves. Tannins are mainly confined to the seed hull.

Cultivars with high tannin content are more tolerant to frost around harvest time (Cho *et al.*, 2019). Moreover, zero-tannin varieties seem to be more susceptible to soil-borne diseases. Plants with a higher tannin content have better chances under

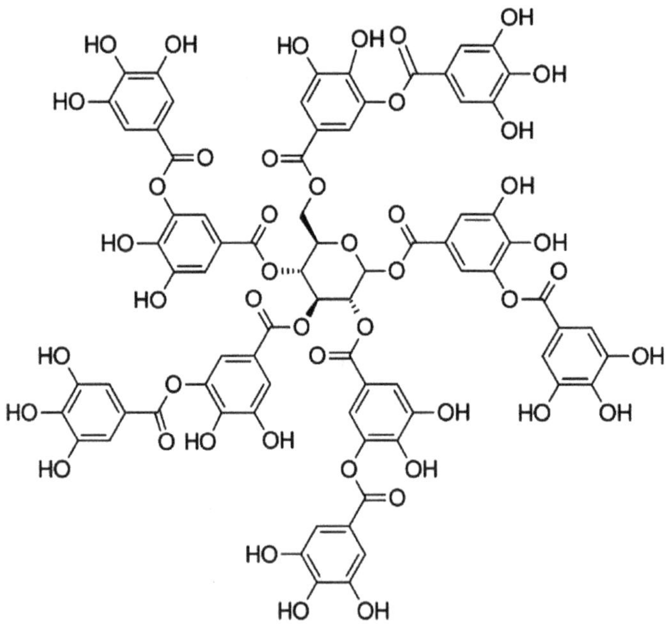

FIGURE 2.3.6-1 Chemical structure of tannic acid
EN:USER_TALK:RONHJONES, PUBLIC DOMAIN, VIA WIKIMEDIA COMMONS

TABLE 2.3.6-1 Tannin content in various legume species

Species	Cultivar	Tannin mg / kg DM			No. of cultivars	Authors
		Mean	Min.	Max.		
Soybeans (*Glycine max L.*)		1.38	0.88	2.06	20	Malenčić *et al.*, 2007
Field beans (*Vicia faba L.*)	Mulitcolour	6.60	5.50	10.40	6	Duc *et al.*, 1999
	White	0.10	0.10	0.20	6	
Grain peas (*Pisum sativum L.*)	Mulitcolour	5.49	0.55	7.45	36	Bastianelli *et al.*, 1998
	White	0.07	0.04	0.14	12	
Sweet lupins[1] (*Lupinus sp.*)	*luteus*	0.02	0.02	0.03	2	Lampart-Szczapaa *et al.*, 2003
	albus	0.05	0.03	0.08	2	
	angustifolius	0.01	0.01	0.02	2	

1 Lupin meal.

difficult growing conditions. Therefore, higher tannin levels have to be anticipated when legumes grew under adverse conditions. In legumes, the highest concentrations of tannins are found in peas (*Pisum sativum L.*) and faba beans (*Vicia faba L.*), predominantly in varieties with coloured flowers and seed coats. White-flowered cultivars with bright coats have a lower tannin content than colour-flowered varieties with darker seeds. Lupins are virtually tannin free (table 2.3.6-1).

Naturally occurring food legume tannins are reported to interact with proteins (both enzyme and nonenzyme proteins) to form tannin-protein complexes, resulting in inactivation of digestive enzymes and protein insolubility. Thus they reduce the nutrient digestibility of dry matter, energy and especially of proteins. Cultivars with a high tannin level usually have a lower energy content. On the other hand, while overall protein digestibility is reduced the amount of rumen bypass protein is enhanced. Studies revealed that feeding a tannin rich faba bean variety did not lower milk yield or milk protein content while stabilizing the ruminal pH and significantly reducing milk urea content (Bissinger *et al.*, 2007).

The mayor antinutritive effect of tannin rich diets is the lower feed intake. Tannins have a bitter taste to which especially pigs and poultry are sensitive. Cows are less susceptible and studies showed that feed intake was not reduced when integrating a faba bean variety with a high tannin content in a dairy cow diet (Bissinger *et al.*, 2004). Reduction of tannins can either be achieved by dehulling seed legumes, since the major part of the component is located in the seed coat. Heating is not effective enough in reducing the tannin concentration. Better, however, is the use of a cultivar with a genetically low tannin content. The tannin content is more related to the cultivar than to growing conditions.

On the other hand, there may also be beneficial aspects of tannins in the diet. There is some evidence that tannins reduce ruminal methane production, at least when supplemented as purified tannin extracts (Hess *et al.* 2006).

2.3.7 *Proteins*

2.3.7.1 Lectins

Of the total protein of legume seeds, up to 10% are lectins, mainly located in the cotyledons. Only small amounts are detected in hull fractions. Lectins have a bitter taste. This protects the plant against consumption by predatory invertebrates and higher animals. Generally, they are odourless, resistant to dry temperatures and to low pH. Many lectins are resistant to inactivation by dry heat but are heat sensitive in the presence of moisture. Their heat stability varies between plant species. Differences in the lectin content between varieties have been found in beans, some of which have no lectins at all.

Lectins are sugar-binding proteins which agglutinate cells and precipitate glycoconjugates. The binding to monosaccharides or oligosaccharides is reversible and highly specific. Lectins adhere to the glycoproteins present on the inner lining of the small intestine and inhibit absorption of feed metabolites. Thus, they reduce feed efficiency. They can cause severe intestinal damage and disrupt digestion by causing epithelial lesions within the intestine. Thus they can interfere with absorption and transportation of nutrients during digestion and cause nutrient deficiencies as well severe scouring, especially in young animals. Therefore, a familiarisation period should be taken into account when using soybean meal in young animal feed for the first time.

Lectins are also known as phytohaemagglutinins, due to their ability to bind to carbohydrates on the membrane of erythrocytes and agglutinate them. There are more than 100 different structures of legume lectins reported. Due to differences in analytical methods, haemagglutinating activities are difficult to compare between studies and are easier to interpret within one publication. In most studies, highest amounts of haemagglutinating activities are found in defatted soybeans, followed by peas (table 2.3.7-1). Faba beans and lupins have low activities and did not cause haemagglutination with rabbit red blood cells. As with other ANFs, the activity varies between cultivars of the same species. Moreover, animal studies gave evidence that the toxicity of lectins does not completely correspond to its haemagglutinating activity. Especially those in soybeans seem to be more toxic than in other legumes.

Dietary lectins also act as protein antigens which bind to the surface glycoproteins on blood cells, thus activating IgG and IgM antibodies and causing food allergies and changes in gut immune function. Further effects of feeding lectin containing diets include the lower production of endocrine cells and gut hormones and the interference with the bacterial ecology in the gut lumen.

On the other hand, lectins exhibit antimicrobial and some also antifungal effects. Their antibacterial activity results from their interaction with a wide variety

TABLE 2.3.7-1 Haemagglutinating effects in various legume species (units/mg DM)

Species	Cultivar	Haemagglutinating effects	
		Valdebouze *et al.*, 1980[1]	Elkowicz and Sosulski, 1982
Soybeans (*Glycine max L.*), defatted		1.600–3.200	77.4
Faba beans (*Vicia faba L.*)		25–100	10.6
Grain peas (*Pisum sativum L.*)		100–400	15.1
Sweet lupins (*Lupinus sp.*)	*luteus*	0.05	*
	albus	0.1	0
	angustifolius	< 0.05	*

1 Rabbit trypsinized red blood cell agglutination test.
* No haemagglutinating activity detected.

of complex carbohydrates of the bacterial cell wall. More research into these properties is needed.

2.3.7.2 Protease inhibitors

Protease inhibitors are proteins with specific antitrypsin and – to a lesser extent – antichymotrypsin activity. Both activities are highly correlated. Generally, trysin inhibitor activity (TIA) is used as measure to determine the overall protease inhibitor activity.

Protease inhibitors decrease the digestibility of proteins by forming indigestible complexes with dietary protein and cause pancreatic hypertrophy. This impairs protein digestion by monogastric animals and some young ruminants, even in the presence of high levels of digestive enzymes. Due to their protein nature, protease inhibitors may easily be denatured by technological treatments like extrusion, infrared radiation, micronizing, autoclaving or flaking, provided the temperature of the treatment is above 80 °C (Gatel, 1992). Toasted soybeans had only 5% of TIA compared to the raw variety, although some residual activity may still remain in the commercially produced products. The antinutritive activity of protease inhibitors is associated with growth inhibition and pancreatic hypertrophy.

Protease inhibitors are found in raw legume seeds. They are mainly located in the cotyledons whose TIA is about 13 times higher than that of the hulls. Most grain legumes contain rather low TIA (Table 2.3.7-2). However, concentrations in raw soybeans are usually high enough to reduce protein digestibility. Soybean meal may as well contain higher amounts of TIA, depending on the production process. The level of TIA also depends on the cultivar of a legume species and on the growing conditions. In contrast to tannins, TIA is much lower in colour-flowered faba bean cultivars compared to white-flowered.

However, in animal studies up to 3.2 mg TIA/g diet did not affect pancreatic nitrogen or protein secretion or pancreatic chymotrypsin activity in young pigs

TABLE 2.3.7-2 Trypsin inhibitor activity (TIA, mg TI/g CP) in various legume species

Species	Variety	TI[1] mg/g CP			No. of cultivars
		Mean	Min.	Max.	
Soybean meal (*Glycine max L.*)		5.8	*	*	1
Faba beans (*Vicia faba L.*)	Colour-flowered	< 0.2	< 0.2	1.4	4
	White-flowered	3.6	3.3	3.9	2
Grain peas (*Pisum sativum L.*)	White-flowered	3.0	< 0.2	5.0	6
Sweet lupins (*Lupinus sp.*)	*luteus*	< 0.2	*	*	1
	angustifolius	< 0.2	< 0.2	2.9	5

1 TI: trypsin inhibitor.
* No data.
JEZIERNY (2009)

TABLE 2.3.8-1 Overview of antinutritional factors (ANF) in grain legumes

Species	Variety	Alkaloids μg/g[12]	Phytate g/kg	Oligo sacchar g/kg[1]
Soybean meal (*Glycine max L.*)		*	12.7[13]	*
Faba beans (*Vicia faba L.*)	Colour	*	1.7–21.25[11]	37.9–46.6
	White			
Grain peas (*Pisum sativum L.*)	Colour	*	7.3[2]	84.9
	White			83.5
Sweet lupins (*Lupinus ssp*)	*albus*	143–226	8.2[15]	71[14]
	luteus	500–895	*	110[14]
	angusti folius	44–2120 median: 138[12]	*	71[14]

* No data.; 1 Gdala and Buraczewska, 1997.; 2 Nikopopoulou *et al.*, 2007, modified.; 3 Elango *et al.*, 202
4 Malenčić *et al.*, 2007.; 5 Duc *et al.*, 1999.; 6 Bastianelli *et al.*, 1998.; 7 Lampart-Szczapaa *et al.*, 2003.;
8 Valdebouze *et al.*, 1980.; 9 Jezierny, 2009.; 10 Jezierny *et al.*, 2010.; 11 Meng *et al.*, 2021.; 12 Reinhard *et*

(Gabert *et al.*, 1996). Similarly, according to Batterham *et al.* (1993), growing pigs may tolerate dietary levels of 4.7 mg TIA/g, without significant negative effects on performance criteria such as growth rate, feed intake or feed conversion rate. Huisman and Tolman (2001), however, recommended a lower level for fattening pigs. According to their studies, the TIA level should not exceed a maximum of 0.5 mg/g in the diet.

2.3.8 *Overview of ANFs in grain legumes*

Table 2.3.8-1 gives an overview of the analysed levels of ANFs found in different legume species. More recently, alkaloid content has only been described in lupin varieties and the data given all refer to sweet lupins. There is a wide range between varieties, with *L. albus* having the lowest levels. Considering the threshold of 0.05% of alkaloids for feeding purposes, white lupins seem to be a rather safe feed ingredient. The situation is different for the other two varieties. In *L. luteus* the lowest reported alkaloid content in four sweet lupin samples was as high as 500 µg/g. The range was highest in *L. angustifolius*, although with a low median in a study with 18 samples. It is therefore important to know the average alkaloid content of the cultivar when including *L. angustifolius* in the diet. The alkaloid content of soybeans, peas and faba beans has not been analysed in recent European, American or Australian publications. It is likely that alkaloids play a minor role in the other legumes. However, the content may be increased under stressful growing conditions (Weber and Krieg, 2023).

The phytic acid content is highest in soybeans and lowest in faba beans. Phytases should always be added to diets containing soybean meal in order to improve the utilisation of dietary P and reduce P excretion.

…actosides[3] …g	Tannin mg/kg	Haemag glutinin units/mg[8]	TIA mg/g CP[9]	Saponin g/kg[10]
2.5	0.88–2.06[4]	1600–3200	5.8	49.4
…7	5.5–7.4[5]	25–100	< 0.2–1.4	25.5–39.3
	0.1–0.2[5]		3.3–3.9	16–18
…6	0.55–7.45[6]	100–400	*	*
	0.04–0.14[6]		< 0.2–5.0	
…7	0.02–0.03[7]	0.05	*	< 0.012
0.2	0.03–0.08[7]	0.1	< 0.2	0.055
	0.01–0.02[7]	< 0.05	< 0.2–2.9	0.379–0.740

…13 Egounlety and Aworh, 2003, modified.; 14 Martínez-Villaluenga *et al.*, 2006, modified.;
…owitz and Sosulsi, 1982.

The effects of oligosaccharides are controversial. On the one hand, they are a good source of nutrients for the gut microbiota and thus promote gut health. On the other hand, α-galactosides cause flatulence, which is uncomfortable to harmful for the organism. Oligosaccharides are the most abundant in peas and they are the main ANF in peas. But especially the raffinose type oligosaccharides, which cause flatulence, can be high in any legume.

Feeding restrictions caused by tannins are only high in colour-flowered peas and beans. They limit feed intake due to their bitter taste. At least in colour-flowered faba beans, saponins can increase poor acceptance and a reduction in feed intake.

Soybeans have been grown for food and feed for a long time and are generally low in ANF. However, they have the highest level of haemagglutinin units and therefore young animals should have an adaptation period before being fed large amounts. They are also high in trypsin inhibiting activity and should be heat treated before feeding.

References

Bastianelli D., Grosjean F., Peyronnet C., Duparque M., Regnier J.M., 1998: Feeding value of pea (*Pisum sativum L.*). 1. Chemical composition of different categories of pea. Anim. Sci., 67, 609–619.

Batterham E.S., Saini H.S., Andersen L.M., Baigent R.D., 1993: Tolerance of growing pigs to trypsin and chymotrypsin inhibitors in chickpeas (*Cicer arietinum*) and pigeonpeas (*Cajanus cajan*). J. Sci. Food agric., 61, 211–216.

Bissinger C., Schneider K., Steingass H., 2007: Untersuchungen an Milchkühen zur Erhöhung der nXP-Versorgung bei Einsatz von Ackerbohnen, Erbsen und Lupinen. Heimische Körnerleguminosen mit geschütztem Protein in der Milchviehfütterung. UFOP-Schriften, Heft 33, 21–27.

Bissinger C., Steingass H., Drochner W., 2004: Steigerung des Gehaltes an nutzbarem Protein bei Körnerleguminosen mittels ökologisch konformer technischer Bearbeitungsverfahren zur Förderung von Gesundheit und Leistungsfähigkeit von Hochleistungskühen im ökologischen Landbau. BÖL-Bericht-ID 8661.

Chitra U., Vimala V., Singh U., Geervani P., 1995: Variability in phytic acid content and protein digestibility of grain legumes. Plant foods for human nutrition, 47, 163–172.

Cho M., Smit M.N., He L., Kopmels F.C., Beltranena E., 2019: Effect of feeding zero- or high-tannin faba bean cultivars and dehulling on growth performance, carcass traits and yield of saleable cuts of broiler chickens. J. Appl. Poult. Res., 28, 1305–1323.

Crépon K., Marget P., Peyronnet C., Carrouee B., Arese P., Duc G., 2010: Nutritional value of faba bean (*Vicia faba L.*) seeds for feed and food. Field crops research, 115, 329–339.

Duc G., Marget P., Esnault R., Guen J. Le., Bastianelli D., 1999: Genetic variability for feeding value of faba bean seeds (*Vicia faba L.*): comparative chemical composition of isogenics involving zero-tannin and zero-vicine genes. Journal of agricultural science, 133, 185–196.

Egounlety E., Aworh O.C., 2003: Effect of soaking, dehulling, cooking and fermentation with Rhizopus oligosporus on the oligosaccharides, trypsin inhibitor, phytic acid and tannins of soybean (*Glycine max Merr.*), cowpea (*Vigna unguiculata L. Walp*) and groundbean (*Macrotyloma geocarpa Harms*). Journal of food engineering, 56, 249–254. doi:10.1016/s0260-8774(02)00262-5.

Elango D., Rajendran K., Van Der Laan L., Sebastiar S., Raigne J., Thaiparambil N.A., El Haddad N., Raja B., Wang W., Ferela A., Chiteri K.O., Thudi M., Varshney R.K., Chopra S., Singh A., Singh A.K., 2022: Raffinose family oligosaccharides: friend or foe for human and plant health? Front. Plant Sci., 13, 829118. doi:10.3389/fpls.2022.829118.

Elkowitcz K., Sosulski F.W., 1982: Antinutritive factors in eleven legumes and their air-classified protein and starch fractions. Journal of food science, 47, 1301–1304.

Gabert V.M., Sauer W.C., Li S., Fan M.Z., Rademacher M., 1996: Exocrine pancreatic secretions in young pigs fed diets containing faba beans (*Vicia faba L.*) and peas (*Pisum sativum L.*): nitrogen, protein and enzyme secretions. J. Sci. Food agric., 70, 247–255.

Gatel F., 1992: Protein quality of legume seeds for monogastric animals. 1st Europ. Conf. on grain legumes, Angers, France, 461–473.

Gdala J., Buraczewska L., 1997: Chemical composition and carbohydrate content of several varieties of faba bean and pea seeds. J. Anim. Feed sci., 6, 123–135.

Hess H.D., Tiemann T.T., Noto F., Carulla J.E., Kreuzer M., 2006: Strategic use of tannins as means to limit methane emission from ruminant livestock. International congress series, 1293, 164–167.

Huisman J., Tolman G.H., 2001: Antinutritional factors in the plant proteins of diets for non-ruminants. Editors: Garnsworthy, P.C., Wiseman, J., 2001: Recent developments in pig nutrition, Nottingham University Press, Nottingham, 3, 261–322.

Jezierny D., 2009: In vivo and in vitro studies with growing pigs on standardised ileal amino acid digestibilities in grain legumes. Ph.D. thesis. University of Hohenheim, Stuttgart, Cuvillier Verlag Göttingen, Germany.

Jezierny D., Mosenthin R., Bauer E., 2010: The use of grain legumes as a protein source in pig nutrition: a review. Animal feed science and technology, 157, 111–128.

Krížová L., Dadáková K., Kašparovská J., Kašparovský T., 2019: Isoflavones. Molecules, 24, 1076–1104.

Lampart-Szczapaa E., Korczakb J., Nogala-Kaluckaa M., Zawirska-Wojtasiak R., 2003: Antioxidant properties of lupin seed products. Food chemistry, 83, 279–285.

Lessire M., Gallo V., Prato M., Akide-Ndunge O., Mandili G., Marget P., Arese P., Duc G., 2017: Effects of faba beans with different concentrations of vicine and convicine on egg production, egg quality and red blood cells in laying hens. Animal, 11, 1270–1278.

Malenčić D., Popović M., Miladinović J., 2007: Phenolic content and antioxidant properties of soybean (*Glycine max* (*L.*) Merr.) seeds. Molecules, 12, 576–581.

Martínez-Villaluenga C., Frías J., Vidal-Valverde C., 2006: Functional lupin seeds (*Lupinus albus L.* and *Lupinus luteus L.*) after extraction of α-galactosides. Food chem., 98, 291–299.

Meng Z., Liu Q., Zhang Y., Chen J., Sun Z., Ren C., Zhang Z., Cheng X., Huang Y., 2021: Nutritive value of faba bean (*Vicia faba L.*) as a feedstuff resource in livestock nutrition: a review. Food sci. Nutr., 9, 5244–5262.

Muduuli D., Marquardt R.R., Günter W., 1982: Effect of dietary vicine and vitamin E supplementation on the productive performance of growing and laying chickens. Br. J. Nutr., 41, 53–60.

Nikolopoulou D., Grigorakis K., Stasini M., Alexis M.N., Iliadis K., 2007: Differences in chemical composition of field pea (*Pisum sativum L.*) cultivars: effects of cultivation area and year. Food Chemistry, 103, 847–852.

Reinhard H., Rupp, H. Sager F., Streule M., Zoller O., 2006: Quinolizidine alkaloids and phomopsins in lupin seeds and lupin containing food. Journal of chromatography A, 1112, 353–360.

Valdebouze P., Bergeron T., Gaborit T., Delort-Laval J., 1980: Content and distribution of trypsin inhibitors and hemagglutinins in some legume seeds. Can. J. Plant Sci., 60, 695–701.

Weber M., Krieg J., 2024: Einsatz unterschiedlicher Mengen an weißen Lupinen in der Schweinemast. Proteinmarkt, 05.06.2024.

2.4 Grain peas (*Pisum sativum L.*)

Mechthild Freitag

After soybeans, grain peas are the most important legume in the European Union. They are used for human as well as for animal nutrition. Because of their protein content of usually more than 20% they play a major role in organic food and feed as

FIGURE 2.4-1
Young peas with pods

FIGURE 2.4-2
White-flowered peas

an alternative to soybean meal. Wild peas have already been part of the diet of the early collective economy. They belong to the most ancient cultivated plants of the Near East domestication centre and spread from there to Europe.

2.4.1 *Ingredients*

2.4.1.1 Nutrients and antinutritive Factors (ANF)

2.4.1.1.1 *Nutrients*

For feeding purposes, there are two varieties of major importance. They can be distinguished by the colour of their flowers. Most prominent are the white-flowered summer peas compared to colour-flowered winter cultivars. The latter have a more or less bitter taste due to a higher content of tannins. This can restrict their inclusion in diets. However, due to higher yields in adverse growing conditions they are important in organic farming.

White and colour-flowered varieties do not differ in their nutrient content. Both have an average of about 22% protein and 50% starch (DM, table 2.4.1.1-1). Thus, they are a protein as well as an energy source. Lipid and fibre (NDF, ADF) concentrations are relatively low. Variations in the nutrient content, especially in the protein level, are not only due to differences in cultivars but more important to environmental conditions during growth: temperature, humidity, fertilization and soil composition.

TABLE 2.4.1.1-1 Nutrients (g/kg DM) of grain peas (white- and colour-flowered)

Ingredient	Peas (Summer, white-flowered)			Peas (Winter, colour-flowered)	
DM	880	1000	min.–max. 1000	880	1000
CP	200	227	171–295	195	222
CL	13	15	11–23	17	19
CF	57	65	57–80	58	66
$aNDF_{om}$	100	114	91–136	123[1]	140[1]
ADF_{om}	70	80	68–91	83[1]	94[1]
STA	430	489	398	447	508
SUG	40	45	23–68	47	53
NFE	577	656	*	587	667
ASH	33	38	28–57	23	27

1 NDF, ADF.
* No data.

BELLOF AND FREITAG, 2021

FIGURE 2.4.1.1-1
Pea pod of white-flowered peas

FIGURE 2.4.1.1-2
Pea pod with seeds of white-flowered peas

FIGURE 2.4.1.1-3
Seeds of white-flowered peas

The composition of starch is different from that of cereals. Pea starch has a high content of amylose compared to amylopectin. Amylose is more resistant to pancreatic amylase activity. Therefore, the starch is more slowly digested than that of cereals (Kirn *et al.*, 2022).

2.4.1.1.2 *Carbohydrates*

Even though peas are considered a protein feed in animal nutrition, the main components are carbohydrates. Besides starch – the main nutrient – different fibre fractions play a major role. The sugar content is low. Total starch amounts to a range of about 400–570 g per kg DM. It mainly consists of digestible (enzyme susceptible) and rumen degradable starch (table 2.4.1.1-2). There are no significant differences in this respect between white- and colour-flowered varieties but the processing method has a pronounced effect on rumen degradability.

TABLE 2.4.1.1-2 Ruminal starch degradation of peas (*Pisum sativum L.*)

Treatment	Immediately degradable %	Degradability rate %
Peas	46.2	84.6
Ground peas	49.4	91.2
Cracked peas	41.1	89.3
Toasted peas	27.4	66.0
Extruded peas	69.9	96.5

OFFNER *ET AL.*, 2003, MODIFIED

TABLE 2.4.1.1-3 Carbohydrate content (g/kg DM) of white-flowered peas (*Pisum sativum L.*)

Constituent	Content
Total starch	369–490
Resistant starch	21–63
Amylose	207–337
Total dietary fibre	140–126
Insoluble fibre	100–150
Soluble fibre	20–90
Soluble sugars	12–24

DAHL *ET AL.*, 2012

According to a study by Gdala and Buraczewska (1997) the carbohydrate content in different cultivars is quite similar, with a starch digestibility of 80 to 90%. A fraction of about 20 to 70 g is enzyme resistant (Gdala and Buraczewska, 1997; Dahl *et al.*, 2012). Compared with cereals, peas contain higher levels of enzyme-resistant and slowly digestible starch.

The fibre content of peas is lower than that of other legumes. Dietary fibre in peas arises from both, the hull and the cotyledon. The hull contains largely water-insoluble polysaccharides, primarily cellulose, whereas the cotyledon fibre consists of polysaccharides with various degrees of solubility, including hemicelluloses and pectins, along with cellulose (Dahl *et al.*, 2012). However, the fibre content varies in relation to cultivars and growing conditions. The total amount of fibre does not differ between white- and colour-flowered cultivars, but NDF, ADF and cellulose content are significantly higher in the colour-flowered varieties (table 2.4.1.1-4).

Approximately 18% of DM consists of non-starch polysaccharides (NSP). Of these, approximately 50% is glucose, followed by arabinose with about 20% and uronic acids with 12%. All other NSP are less than 10% (2.4.1.1-5). There are no significant differences between white- and colour-flowered peas.

Peas contain large amounts of oligosaccharides that are rapidly fermented in the rumen. In both cultivars discussed above, about 70% of these are of the raffinose type (raffinose, stachyose, verbascose) and 30% are sucrose (table 2.4.1.1-6). Moreover, cell-wall substances are associated with significant amounts of pectic substances which, for analytical reasons, are not accounted for in the NDF residue.

TABLE 2.4.1.1-4 Starch and fibre content (g/kg DM) in white- and colour-flowered peas ($\overline{x} \pm$ SEM of 7 white- and 2 colour-flowered cultivars)

Parameter	White-flowered	Colour-flowered	SEM
Total starch	461	459	8.24
Enzyme susceptible starch	419	390	5.19
NSP	182	186	4.79
NDF	118^{b}	140^{a}	2.89
ADF	78^{b}	94^{a}	1.65
CF	58	65	1.02
ADL	3	4	0.30
Hemicelluloses	37	45	2.67
Cellulose	75^{b}	90^{a}	2.11

a,b $p \leq 0.05$.

GDALA AND BURACZEWSKA, 1997, MODIFIED

TABLE 2.4.1.1-5 Composition and content (g/kg DM) of non-starch polysaccharides in white- and colour-flowered peas ($\overline{x} \pm$ SEM of 7 white- and 2 colour-flowered cultivars)

Parameter	White-flowered	Colour-flowered	SEM
Total NSP	182.0	186.4	4.79
Glucose	86.3	92.9	2.80
Arabinose	36.7	37.0	1.38
Uronic acid	23.1	23.5	0.53
Xylose	16.0	14.3	0.69
Galactose	14.7	14.4	0.93
Rhamnose	2.4	2.2	0.10
Mannose	2.1	1.5	0.21
Fructose	0.7	0.7	0.04

GDALA AND BURACZEWSKA, 1997, MODIFIED

TABLE 2.4.1.1-6 Composition and content (g/kg DM) of oligosaccharides in white- and colour-flowered peas ($\overline{x} \pm$ SEM of 7 white- and 2 colour-flowered cultivars)

Parameter	White-flowered	Coloured-flowered	SEM
Total oligosaccharides	83.5	84.9	1.42
Sucrose	26.4	23.4	1.07
Total α-galactosides	57.1	61.5	0.97
Raffinose	7.3	6.3	0.44
Stachyose	24.9	25.3	1.54
Verbascose	24.9	29.9	1.62

GDALA AND BURACZEWSKA, 1997, MODIFIED

2.4.1.1.3 *Antinutritive factors (ANF)*

For more extensive descriptions of ANF in legumes refer to section 2.3.

Compared to other legumes, most pea cultivars are low in ANF (table 2.4.1.1-6). The mean values of the total amount of non-starch polysaccharides (NSP) are about 15 to 19%. NSP content is positively related to protein content but negatively correlated with starch (Nikolopoulou *et al.*, 2007). Oligosaccharides may cause flatulation. Their content is lower than in soybeans and comparable to faba beans. Trypsin inhibitor activities (TIA) are also very low, compared to soybeans.

TABLE 2.4.1.1-7 Variation of ANF content in peas (% DM) in three cultivars and three growing locations

Item	Mean	Min.	Max.	Effect of	
				Location	Year
Tannins	0.66	0.45	0.92	**	**
Phytic acid	0.73	0.21	1.26	**	*
NSP	16.3	14.4	19.5	**	*
Raffinose series oligosaccharides	6.2	5.8	7.3	–	**

* p < 0,05.
** p < 0,01.
– p > 0,05.
NIKOPOPOULOU *ET AL.*, 2007

Therefore, heating is not essential to gain optimum protein digestibility. Usually, TIA is not high enough to reduce pancreatic trypsin and chymotrypsin activities.

Some pea cultivars may contain saponins. They impart a bitter taste to the feed and thus reduce palatability and feed intake. The tannin content of peas is quite variable and generally related to flower colour rather than to seed colour. White-flowered peas (*P. sativum hortense L.*) almost always contain negligible amounts whereas colour-flowered peas (*P. sativum arvense L.*) can contain 5–10 times more tannins and may cause a reduction in feed intake.

2.4.1.2 Minerals and vitamins

Mineral content of peas is similar to cereals (table 2.4.1.2-1): rich in phosphorus (P) (3.5–5 g/kg DM) and low in calcium (Ca) (0.6–2.0 g/kg DM). Tannins may restrict the mineral digestibility. In general, P digestibility is approximately 50% for pigs and 60% for poultry. Compared to wheat, peas contain more potassium (K) (13 g/kg DM) and twice as much ferrum (Fe) (106 g/kg DM) as cereals. All in all, there are great variations in different cultivars. Fe content can be increased 3.5 fold due to an uncontrolled hyperaccumulation into vegetative tissues and the ability to transport excess Fe to seeds (Wang *et al.*, 2003). However, mineral digestibility depends on the phytate concentration of the cultivar. Intestinal mineral digestibility of Ca, P, Zn and Fe can be increased by the addition of phytases to the feed or by fermentation. The P digestibility values for pigs and poultry are listed in table 2.4.1.2-1.

TABLE 2.4.1.2-1 Average mineral content (in DM) of peas

Item		White-flowered peas
Ca	g	1.1
P	g	4.7
DCP_{Pig}[1]	g	2.2
$DCP_{Poultry}$	g	3.0
Mg	g	1.5
Na	g	0.2
K	g	13
Cl	g	0.9
S	g	2.3
Co	mg	0.1
Cu	mg	8.1
Fe	mg	106
J	mg	0.3
Mn	mg	10
Mo	mg	2.31
Zn	mg	37
Se	mg	0.17
DCAB[2]	meq	119
BE[3]	mmol/kg	265

1 Digestible P
2 Dietary cation-anion balance.
3 Base excess.
BELLOF AND FREITAG, 2021

The content of trace minerals, e.g. Co, Cu, Fe, J, Mn, Mo, Zn and Se varies considerably, depending on the cultivar, mineral content of the soil and their biological availability and to weather conditions during growth. Therefore, before adding large amounts of peas in a diet an analysis of their main nutrients should be conducted.

Peas are rich in vitamin A and Cholin and low in vitamin E (Table 2.4.1.2-2). According to a review by Savage and Deo (1989), the vitamin C content varies considerably. The authors stated concentrations from 22 to 295 mg/kg DM.

TABLE 2.4.1.2-2 Vitamin content of white-flowered peas

Vitamin		Content
Vitamin A	mg/kg	500
Vitamin E	mg/kg	1.0
Thiamin (B_1)	mg/kg	8.1
Riboflavin (B_2)	mg/kg	2.4
Niacin	mg/kg	32
B_6	mg/kg	1.9
B_{12}	mg/kg	0
Pantothenic acid	mg/kg	19
Cholin	mg/kg	685
Biotin	µg/kg	200
Folic acid	mg/kg	4.2

BELLOF AND FREITAG, 2021

2.4.2 *Feed value*

FIGURE 2.4.2-1
Seeds of white-flowered peas

2.4.2.1 Protein

In legumes, proteins mainly consist of storage proteins (globulins, albumins, glutelins) in the embryo to supply amino acids during germination. Other proteins, such as protease and amylase inhibitors or lectins are only present in small amounts. Despite their low concentrations they can restrict pea utilization in feed.

In ruminants, protein quality depends on the nitrogen content and the speed and amount of rumen degradation. In relation to the protein content the amount of rumen stable protein or protein availability in the small intestine does not differ greatly from soybean or rapeseed meal. However, peas only deliver about half the amount of protein to the small intestine per kg compared to soybean meal. About twice as much is fermented in the rumen with only 15% bypass protein (white-flowered peas). Thus, in many diets for high performing dairy cows a second source of protein will have to be included. Rape seed, with its low content of lysine and high amount of methionine, is an ideal complementary protein source.

On the other hand, ensiling and/or hydrothermal treatment of peas results in a reduced rate of ruminal fermentation. Depending on the rumen fermentation rate, hydrothermal treatment of peas considerably increased the amount of available protein in the small intestine in vitro. Moreover, feed with toasted peas was highly palatable for dairy cows. Inclusion of 4 kg peas per animal and day is possible without negative effects (Preissinger *et al.*, 2007).

To reduce harvest losses peas can be threshed at a dry matter content of 50% and then conserved by ensiling. This process also augments the amount of rumen bypass protein (Zeyner *et al.*, 2021) without reducing the nutrient content. The treatment reduced protein solubility (table 2.4.2.1-1) and N excretion (table 2.4.2.1-2). Substitution of rapeseed meal by ensiled and toasted peas had no effect on feed intake and milk yield (ECM). However, it enhanced milk fat and reduced milk protein content. The overall N excretion via faeces and urine was markedly reduced (Steinhöfel and Kuhnitzsch, 2021).

For monogastric animals, peas are a source of protein as well as of energy. Digestibility of crude protein varies from 80 to 91% for white-flowered peas and from 71 to 84% for colour-flowered peas. This difference probably results from the presence of tannins in the coloured varieties (Gatel and Grosjean, 1990). The apparent ileal digestibility does not vary between cultivars with low or high TIA concentration (Le Guen *et al.*, 1995). Peas contain a relatively high amount of lysine and valine but are low in methionine, cysteine, threonine and tryptophan, all of them

TABLE 2.4.2.1-1 Effects of ensiling and toasting of peas on protein solubility

Parameter		Untreated	Ensiled	Ensiled and toasted*
DM (g/kg OM)		784	757	927
CP (g/kg product)		189	189	196
Protein solubility (% of CP)		75.5	42.7	17.3
CNCPS protein fractions	A	6.9	9.3	12.5
(% of CP)	B1	68.7	33.4	4.82
	B2	23.6	54.8	47.0
	B3	0.8	6.0	31.8
	C	0.5	0.6	3.9
Rumen bypass CP (% of CP)		4	9	32

* Max. seed temperature 70 °C.
KUHNITZSCH AND STEINHÖFEL, 2021

TABLE 2.4.2.1-2 Effect of treatment of peas on N excretion

Parameter		Without treatment	Ensiled and toasted
Total N in urine	g/l	8.6[a]	13.3[b]
Uria in urine	g/l	9.0[a]	14.3[b]
Total N in faeces	g/kg DM	31.4[a]	34.8[b]
CP in faeces	g/kg DM	196[a]	217[b]

a,b $p \leq 0.05$.
STEINHÖFEL AND KUHNITZSCH, 2021

TABLE 2.4.2.1-3 Amino acid content and digestibility for pigs and poultry of white-flowered peas

Amino acid	Content			Digestibility	
	g/kg (88% DM, 200 g CP)	g/100g CP	g/100g CP (Min–max)	Pig %	Poultry %
Lys	15.0	7.5	6.9–8.0	85	88
Met	1.9	1.0	0.9–1.1	77	80
Cys	2.5	1.3	1.1–1.5	68	70
Thr	7.9	4.0	3.5–4.5	76	78
Trp	1.9	1.0	0.8–1.1	68	75
Ile	8.6	4.3	3.4–4.9	82	84
Leu	14.7	7.4	6.3–8.0	80	87
Val	9.7	4.9	3.8–5.5	78	82

BELLOF AND FREITAG, 2021

with a moderate to low intestinal digestibility of about 77% (methionine, threonine) and 68% (cysteine, tryptophan). Therefore, in most diets for monogastric animals, methionine and tryptophan will have to be added. The amounts of essentials amino acids of white-flowered peas and their digestibility for pigs and poultry are listed in table 2.4.2.1-3. The digestibility may vary according to the pea variety and age of animals.

2.4.2.2 Energy

In spite of a high crude fibre level, the energy digestibility is high. This can be explained by the high digestibility of the nitrogen-free extract, crude fibre and crude protein. The digestible energy content of colour-flowered peas appears lower than for white-flowered varieties. This results from the activity of tannins which

TABLE 2.4.2.1-4 Winter and summer peas: apparent digestibility and digestibility coefficient (DC in % of ME) in pig fattening

Ingredient	OM	CP	CL	CF	NFE	MJ ME[2] / kg DM
Summer pea	93.3	100.8	−21.1	124.9	94.3	16.01
Winter pea, untreated	83.6	95.6	−69.8	92.4	88.3	14.58
Winter pea, treated[1]	88.1	98.5	6.8	93.8	91.7	15.20

1 Hydrothermic treatment.
2 ME is calculated with 100% if DC is > 100.
EBERT AND BERK, 2013

inhibit α-amylase and cellulase activity of bacteria in the hindgut. They decrease the apparent digestibility of energy (Gatel and Grosjean, 1990).

Due to the high starch content and overall digestibility of the organic matter of more than 90% for pigs, the energy content of white-flowered peas (summer peas) is similar to that of wheat (table 2.4.2.1-4). Tannins reduce overall intestinal tract digestibility. For colour-flowered winter peas this results in a reduction of OM digestibility of about ten percentage points and therefore in a reduced energy value (table 2.4.2.1-4). Hydrothermic treatment could reduce the negative effects. However, in feeding trials winter peas had no effect on the fattening performance of pigs (Ebert and Berk, 2013).

2.4.3 *Feeding recommendations*

Legumes in general are valuable as protein source and feed material for all domestic animals. However, due to their antinutritive factors acceptable amounts in feed composition vary according to legume species and variety as well as animal species and age. Over the last decades several feeding trials have been carried out to determine the acceptable amount of new pea cultivars.

However, when using peas in compound feed, it must be kept in mind that their seeds are larger and therefore more difficult to dry than cereal grains. Residual moisture above 14%, for example due to overgrowth or uneven ripening, can cause mould growth. Animals reject these peas. Mouldy peas can be a reason for reduced feed intake.

2.4.3.1 Cattle and small ruminants

To assess the feed value of plants for ruminants, the influence on rumination activity and the effects on rumen fermentation must be taken into account in addition to the supply of nutrients. The stimulation of rumination depends on the one hand on the particle size and, on the other hand, on the mechanical resistance to chewing. The latter is estimated by the content of structurally active fibres (NDF, ADF,

FIGURE 2.4.3-1 Grain peas harvest

CF) and their chemical composition (table 2.4.1.1-3). The fibre composition of peas can be compared to that of cereals. In a feed ration they do not contribute to the fibre supply. However, a larger part of the fibres consists of easily fermentable components (tables 2.4.1.1-3 and 2.4.1.1-5) and serves as an energy source for microorganisms in the rumen.

For ruminants, protein quality is mainly determined by the nitrogen content and by the speed and extent of ruminal degradation. The amino acid composition is only relevant in the rumen bypass protein.

The nitrogen content is shown in the crude protein value. The fermentation rate of protein depends on the level of feed intake, which determines the ruminal retention time. The extent of protein conversion in the rumen is also influenced by the protein structure and the incrustation with poorly fermentable fibres. A certain amount of protein passes through the rumen unchanged. This is reflected in the RUP (undegraded protein) value which is even lower in peas than in wheat (15% in white-flowered pea cultivars and 20% in wheat). It can be augmented by ensiling and even more by heating (table 2.4.3.1-1). In situ research into the effects of toasting enhanced the RUP content threefold in peas (Preissinger *et al.*, 2007) and in the ensuing feeding trial a significantly higher total milk yield was recorded. However, RUP enhancement had no positive effect on milk protein and milk urea content (table 2.4.3.1-1). The authors concluded that the effect on milk yield was probably

TABLE 2.4.3.1-1 Effects of toasted peas on milk yield and milk composition

Parameter		Peas	
		Raw	Toasted[1]
Protein content of peas	g/kg	203	206
UDP[2]	g/kg DM	7	21
Microbial protein[2]	g/kg DM	152	152
Feed intake	kg DM/d	21.4	21.4
Rumen pH		6.5[a]	6.3[b]
Milk yield	kg/d	30.6[a]	31.9[b]
Milk fat	%	3.96	3.95
Milk protein	%	3.35	3.32
FECM[3]	kg/d	30.5[a]	31.8[b]
Urea	mg/l	255	258

1 Toasting procedure: Jet sploder at 250 °C air temperature for 1–2 minutes (160 °C in the seed).
2 in situ with estimated rumen passage rate of 8%/h.
3 Fat and protein corrected milk.
a,b $p < 0.05$.
PREISSINGER *ET AL.*, 2007

due to an enhanced energy availability, because the ruminal pH value was lower when cows were fed toasted peas.

Feeding trials with peas as sole protein source have been carried out with 2 to 5 kg of peas per cow per day in the compound feed. No negative effects were detected compared to diets with rapeseed or soybeans or a combination of both. Feed intake was not adversely affected.

Other feeding trials aimed at improving protein availability for ruminants through tannin-rich varieties or heat treatment. Cattle are less susceptible to the bitter taste of tannins and accept tannin rich cultivars (colour-flowered winter varieties). Heat treatment reduces the amount of protein degradation in the rumen and enhances the level of bypass protein. However, as this may speed the ruminal starch fermentation and reduce ruminal pH value heat treated peas may be adverse in compound feed for high yielding cows. Restrictions in the amount of peas in feed rations have to be considered in so far, as total starch and other rapidly rumen fermentable carbohydrates are concerned. Rumen degradable starch and sugar should not exceed 250 g/kg DM in the compound feed for dairy cows.

Peas have only small amounts of rumen bypass protein and bypass starch. Both are required for high yielding cows and may have to be included in the diet by other protein sources.

TABLE 2.4.3.1-2 Recommendations for the inclusion of peas in cattle feed

Cattle category	g/100 kg BW[1]	kg/animal	% of compound feed
Calves (Up to 4 months of age)	–	–	20
Heifers (From 4 months on)	300	–	30
Dairy cows	600	4.0 (650 kg BW)	40
Beef cattle	500	2.0 (450 kg BW)	45
Suckler cows	600	2.0 (300 kg BW) 3.0 (500 kg BW) 4.0 (700 kg BW)	

1 BW: body weight.
HOFFMANN AND STEINHÖFEL, 2023

TABLE 2.4.3.1-3 Recommendations for the inclusion of peas feed for small ruminants

Animal group	g/100 kg BW[1]	g/animal/day	% of feed
Sheep	360	250 (70 kg BW)	20–45 (compound feed)
Lamb fattening (With concentrates)	*	*	20–30 (concentrates)
Lamb fattening (With low protein roughage)	*	*	45 (concentrates)
Lamb fattening (With high protein roughage)	*	*	35 (concentrates)
Dairy goats	250	150 (60 kg BW)	10 (compound feed)
Goatling			10 (compound feed)

1 BW: body weight.
* No data.
BELLOF AND FREITAG, 2021; HOFFMANN AND STEINHÖFEL, 2023

Recommendations for the inclusion of peas in the diet of young cattle are 20% in concentrates, but older cattle will accept peas at higher levels (table 2.4.3.1-2).

Recommendations for beef cattle are up to 2.0 kg per day for younger animals but can be increased to 2.5 kg per day in combination with maize silage. Because of their high starch content, peas can substitute soybeans or rapeseed and cereals in compound feed for beef cattle with average daily weight gain of less than 1200 g.

Fast growing beef cattle with daily weight gain above 1200 g need more bypass protein. Therefore, a combination of peas with soybeans or rapeseed is recommended.

Young sheep lambs accept up to 30% peas in compound feed, whereas young goat lambs are less tolerant. Their compound feed should not exceed 10% of peas (table 2.4.3.1-3). For mature sheep and dairy goats, the concentration of peas in the compound feed depends on the protein content of the roughage and the daily protein demand.

2.4.3.2 Pigs

Feed quality for pigs is mainly determined by the energy content and the amount of digestible protein and digestible essential amino acids (AA). These are predominantly lysine, methionine, threonine, tryptophan and valine. In total, the amount of AAs of peas is only half that of soybeans. But related to the protein content they don't differ much except for methionine, cysteine, and tryptophan. The diets have to be balanced for these AAs. The level of each AA depends on the protein content of the pea, independent of genotype or environment. Amino acid contents are not a fixed proportion of total protein content. They decrease with protein content for a number of essential AAs, particularly lysine and sulphur amino acids (Gatel and Grosjean, 1990).

In new cultivars of white-flowered peas the ANF trypsin inhibitors, lectins, α-galactosides, and tannins are low so that a heat treatment is not necessary. The digestible energy value is similar to soybeans for white-flowered peas (*P. sativum hortense L.*) and slightly lower for dark-flowered pea (*P. sativum arvense L.*) cultivars with higher tannin content.

Several feeding trials have been carried out in pig fattening with conventional and ecological feed to determine to what level peas could be included in compound feed (table 2.4.3.2-1).

In four out of five trials, the inclusion of up to 30% peas in compound feed had no adverse effects on daily weight gain, feed intake, feed efficiency, and carcass quality compared to soybeans or rapeseed. In one trial peas even resulted in significantly higher feed efficiency. Thus, the inclusion of up to 30% of peas to compound feed for pig fattening is recommended.

Piglets seem to be more sensitive to peas than older pigs, especially to raw peas. For young piglets less than 9 kg bodyweight, raw peas should not be fed at levels higher than 10% of the diet. Other authors recommend to exclude peas completely from the diet at that age (Hoffmann and Steinhöfel, 2023). Inclusion of 15–45% of raw peas resulted in decreased feed intake and growth rate (Bertrand *et al.*, 1980; Bengala Freire *et al.*, 1989). However, feeding a diet containing 30 or 45% extruded peas to weaned piglets resulted in nutrient digestibility and weight gain similar to that obtained with the control diet (Bengala Freire *et al.*, 1989). For older piglets weighing 10 kg, peas can generally be incorporated up to 20% of the diet without any adverse effect if the feed is well balanced for all essential AA. Even the inclusion of 30% white-flowered peas did not have adverse effects (Gatel *et al.*, 1989).

TABLE 2.4.3.2-1 Feeding trials to determine the possible integration of peas in feed for pig fattening

Author	Percent of compound feed	Daily gain (g/day)	Feed consumption (kg/day)	Feed efficiency[2]	% lean muscle / index points
Meyer, 2016	0	996	2.71	2.73[a]	1.005
	15/20/25[1]	1017	2.68	2.63[b]	0.995
Heinze *et al.*, 2015a	0	819	2.26	*6	57.7
	15/20/25	822	2.37	*	57.8
Weber, 2016	0	912	2.50[a]	2.76[a]	59.5
	10/15/20	905	2.35[b]	2.61[b]	58.3
	10/20/30	886	2.30[b]	2.61[b]	59.3
	20/20/20	894	2.31[b]	2.59[b]	58.3
Heinze *et al.*, 2015b	0	917	2.53	2.80	57.9
	20/25/30	887	2.46	2.80	57.9
Berk and Ebert, 2016[3]	15/15 (SP)[4]	880	2.93	n.r.	55.6
	15/15 (WP)[5]	905	2.97	n.r.	55.4
	30/30 (WP)	883	2.98	n.r.	54.2

1 Starter / grower / finisher diets.; 2 kg feed per kg weight gain.; 3 Eco-feed.; 4 Summer peas.; 5 Winter peas.; 6 No data.; a,b $p < 0.05$.
WEBER *ET AL.*, 2020

However, this feeding trial was carried out with a dietary protein content of 21%. Diets low in protein may lead to different results.

Feeding sows 16% of white-flowered peas in the gestation diet and 24% during lactation had no adverse effects compared to diets with more cereals and soybean meal (Gatel *et al.*, 1988). To assess long term effects the feeding trial was carried out over at least four reproductive cycles. Diet composition did not affect feed intake of sows or change of body weight. Moreover, no effects could be recorded on weight of piglets at birth and at weaning as well as on piglet mortality rate. In another feeding trial the substitution of 24% wheat by 20% of pea starch and 4.5% of soybean meal in the lactation diet also had no negative effects on daily feed consumption, individual weight, litter weight, blood glucose level after feeding and reproductive performance (Thingnes *et al.*, 2013). Feed consumption was even significantly higher for the pea starch diet, resulting in lower weight loss during the first three weeks of lactation. Only the weaning-to-service interval was longer for the first parity sows offered the pea diet ($p < 0.05$).

TABLE 2.4.3.2-2 Recommendations for the inclusion of white-flowered peas (*Pisum sativum L.*) in compound feed for pigs (% of feed)

Piglets up to 20 kg BW[1]	Piglets more than 20 kg BW	Starter	Finisher	Sow pregnant	Sow lactating
10	20	20	25[2]/40	8	20

1 Body weight.
2 In liquid feed.
WEBER *ET AL.*, 2020

When feeding boars a diet containing 20% of peas, semen quality was not affected (Czarnecki *et al.*, 1988).

In liquid feeding systems the swelling and foaming capacity of pea meal has to be taken into account. In these feeding systems, pea content should be restricted to 25% in the compound feed. Otherwise, pumpability of the liquid feed may be impaired (Hoffmann and Steinhöfel, 2023; Weber *et al.*, 2020).

2.4.3.3 Poultry

Due to the high amylose content of peas, some starch is digested very slowly or not at all during passage through the small intestine. In-vitro analyses indicated that the digestibility rate of pea starch was 21% after 15 minutes and 88% after 120 minutes of enzyme incubation, while wheat starch was digested at rates of 45% and 94%, respectively. Moreover, if poultry was fed a diet of 100% peas, 3.2% starch was analysed in the ileal chymus and only 0.9% if the diet contained no peas (Kirn *et al.*, 2022).

2.4.3.3.1 *Broiler*

For very young chicks (age 1 day to 2 weeks), pea ingredients should be heated and balanced for AAs. A diet in which one half of the protein was provided by ground cull peas was found to closely equal the soybean control for gain and gain/feed ratio, if the peas were autoclaved and the diet supplemented with 0.2% methionine. Extrusion of peas at either 135 °C or 150 °C did not improve performance, but autoclaving at 121 °C for 15 min did. Autoclaving appears to alter or destroy some factor(s) and allows better chick performance. Autoclaving and extrusion at 150 °C completely destroyed the trypsin inhibitor activity present in raw peas, but extrusion at 135 °C did not. All heat treatments decreased the pancreas enlargement observed in chicks fed on raw pea diets. Heating peas has also been shown to increase the metabolisable energy for chicks, but it decreases the lysine content (Owusu-Ansah and McCurdy, 1991).

Peas have a high content of metabolisable energy and, compared to other legumes, are low in antinutritive factors. Therefore, they can be widely used in diets for poultry. In a comprehensive feeding trial boiler chickens were fed starter diets up to 21 days of age and grower diets up to day 35 (Janocha *et al.*, 2022). In starter/grower diets soybean meal was replaced by peas at concentrations of 10 and 20% (group I) and 15 and 25% (group II). In both periods, peas had no adverse effect on body weight and feed conversion rate. Also meat characteristics (dressing percentage, muscularity, fatness) were not different between groups. However, sensory features (colour, flavour, tenderness, juiciness, palatability) were positively affected in both pea groups. These results correspond to those of prior feeding trials. Therefore, for starter/grower chickens an inclusion of 20/25% of peas can be recommended in broiler diets, if essential AAs are balanced. In a previous study, even the inclusion of 30/36% of peas in starter/grower diets had no negative effect on weight gain, feed intake and feed conversion ratio (Farrell *et al.*, 1999). More feeding trials with modern cultivars are needed to assess the maximum inclusion rate for broiler chicken.

2.4.3.3.2 *Layer*

In laying hens from 20 to 54 weeks of age the effects of a replacement of soybean meal by peas was studied. Diets were supplemented with 350 g/kg pea meal, with or without the addition of enzymes to enhance the NSP digestibility (Ciurescu and Pană, 2017). Peas had no significant effect on growth rate, feed intake and feed efficiency or egg production and egg quality traits. Only yolk colour was more intensive in hens fed the pea diet. Enzyme supplementation (Allzyme® Vegpro) significantly improved hen-day egg production and egg mass.

Other authors considered an inclusion of 300 g/kg to be generally accepted by layers.

2.4.3.3.3 *Turkey*

Young turkey poults are sensitive to feeding conditions. However, peas can be introduced in the diet if the amounts are gradually enhanced during the growing period. Recommended amounts of peas are 15% up to 4 weeks of age, 20% from week 5 to 8, 30% from week 9 to 12 and 40% from week 13 to 20. In the finishing period, no more soybean meal is needed. Feeding trials demonstrated that these inclusion rates had no negative effects on weight gain, feed intake (Janusonis *et al.*, 2009), and carcass quality except for the abdominal fat content which was significantly lower in females of the pea groups (Juodka *et al.*, 2016). Soybean meal replacement had no negative effect on the carcass quality of broiler turkeys. Recommendations for the inclusion of peas in turkey diets vary (table 2.4.3.3-2) but soybean meal appears to be unnecessary from week 13. However, diets have to be balanced for essential amino acids.

TABLE 2.4.3.3-1 Recommendations for the inclusion (%) of peas (*Pisum sativum L.*) in compound feed for broiler and layer

Pea variety	Chicken	Pullet	Layer	Brooder	Broiler
White-flowered	30	25*/30	30	30	30
Colour-floured	20	20	20	20	20

* Until week 4; Bellof *et al.*, 2020.
HOFFMANN AND STEINHÖFEL, 2023

TABLE 2.4.3.3-2 Recommendations for the inclusion (%) of peas (*Pisum sativum L.*) in turkey compound feed

Variety	Brooder	Week 1 to 4	Week 5	Week 9	Week 13	Author
White-flowered	15	20	20	20	20	Hoffmann and
Colour-flowered	15	10	30	30	30	Steinhöfel, 2023
White-flowered	*	10/20	30	30	30/25	Bellof *et al.*, 2020, modified
Eiffel	*	15	20	30	40	Janusoris *et al.*, 2009

* No recommendations.

2.4.3.4 Horses

The diet of horses with higher energy and/or protein requirements (growing, lactating, heavy work) is supplemented with concentrates. Main ingredients of these concentrates are oats and soybean meal, main nutrients starch and protein. Both are also the main nutrients in peas and it was tested to what extent oats and soybean meal can be substituted. The substitution of soybean meal is desired for ethical reasons, because most of it is imported into the EU and in Brazil rainforest is cut to grow soya. Oats, on the other hand, are said to cause agitation in some horses. So it may be desirous to substitute as much of these feed ingredients as possible.

In two subsequent years, yearling Quarter Horses were fed three diets with different amounts of oats, peas, and soybean meal. Group a) was fed 80% oats, 0% peas and 13% soybean meal, group b) 57% oats, 29% peas (0.9 kg), and 7.3% soybean meal and group c) 31% oats, 62% peas (1.8 kg), and 0% soybean meal. The daily concentrate ration was divided into two portions. No differences were identified for average daily weight gain or final body weight. There was also no evidence

TABLE 2.4.3.4-1 Recommendations for the inclusion of peas (*Pisum sativum L.*) in equine compound feed

Horse category	In compound feed (%)	Per day (kg)	Per 100 kg BW (kg)	Adult. 600 kg BW (kg)	Author
Foal	12.5	*	*	*	Coenen, 2020
Yearling	60	1.8	*	*	Şentürklü and Landblom, 2015
Adult	20	*	0.25	1.5	Hoffmann and Steinhöfel, 2023

* No recommendation.

of discomfort, colic, hoof laminitis, digestive upset, or feed refusal. It was concluded, that oats can be in parts and soybean meal completely substituted by peas (Şentürklü and Landblom, 2015).

This is supported by recommendations of Coenen (2020). He considers both peas varieties – *Pisum averse L.* and *Pisum sativum L.* – a suitable protein feed, since the tannin content is low in modern cultivars. After adaptation, foals will accept 12.5% peas in their concentrates.

However, for all horses the total amount of starch should not exceed 1 g/kg life weight per feeding. If more energy is needed the total ration has to be divided into several feedings. Moreover, special attention has to be given to the hygienic quality of peas. The rather big seeds need to be carefully dried after harvest. Otherwise they will mould and this can lead to reduced feed intake.

Chicklings (*Lathyrus sativus*) are unsuitable for horses. They contain a poison (Aminopropionitril) that can cause severe damage to nerves and may lead to paralysis of the limps and the larynx (Coenen, 2020).

2.4.3.5 Fishes and shrimps

Monika Weiß, Vanessa Fuchs and Sinem Zeytin

Peas present a high-potential candidate for the replacement of other protein and starch resources in aquafeeds. In this sense, peas can partially replace cereals, which are valuable as binders and energy sources, and protein sources like animal protein or soybean to a certain level (Table 2.4.3.5-1). In practical diets using pea or pea products the sulfur-containing amino acids, mainly methionine, should be balanced, depending on the other feed ingredients. Pea protein concentrates are frequently manufactured, especially for piscivorous fish like salmon, which need high levels of protein (~38% DCP) in their feed (Øverland *et al.*, 2009).

TABLE 2.4.3.5-1 Executive summary of studies testing pea ingredients in aquafeeds for various species

Species	Initial mean weight (g)	Product	Replaced resource	Inclusion rate (%)	Recommendable inclusion rate (%)	Note	Reference s. App. 1
Atlantic salmon (*Salmo salar*)	160 (Seawater)	Protein concentrate (35% u. 50% CP)	FM[a]	20	20		Øverland *et al.* 2009
	2360 (Seawater)	Protein concentrate	FM	35 + 13	13		Penn *et al.* 2011
		Starch	Wheat	8.3 + 16.4		Not exp. tested	Sørensen *et al.* 2011
	546 (seawater)	Protein concentrate	FM	3.5, 7, 10.4 + 15.2	8.90*	Pea as part of mixture	Hansen *et al.* 2011
Rainbow trout (*Oncorhynchus mykiss*)	Exp I: ~277 Exp II: ~190 Exp III: ~437	Flax:pea (F:P) and canola:pea (C:P) mixtures (1:1)	Reference diet	30	30 canola:pea + protease	Canola:pea (C:P) mixtures + protease	Drew *et al.* 2005
	235–825	Pea meal (PM), pea protein concentrate, (PPC)	Other protein ingredients (FM, meat and bone meal, gluten)	0, 75, 150, 225 or 300 g/kg of each test ingredient	No negative effects of PM and PPC	Growth rate dependent on provided nutrients	Collins *et al.* 2012

a Fishmeal.

TABLE 2.4.3.5-1 Executive summary of studies testing pea ingredients in aquafeeds for various species (*cont.*)

Species	Initial mean weight (g)	Product	Replaced resource	Inclusion rate (%)	Recommendable inclusion rate (%)	Note	Reference s. App. 1
	~66	Fixed matrix of organic pea. horse-bean and rapeseed plant protein concentrates	FM protein	0, 5.3, 10.7 + 16 pea protein concentrate	16		Lund *et al.* 2011
	~80	Pea meal	Reference diet	30	30	ADCs dependent on target species	Magalhaes *et al.* 2018
		Pea starch		10 + 20	20	No negative effects on growth, prolonged hyperglycemia	Ji 2017
	~100	Not dehulled. Rough flour extruded	Reference diet	30	Not recommended	Low protein content, not balanced	Burel *et al.* 2000
	~300	Dehulled pea meal pea protein concentrate	Soybean meal	25% dehulled peas 20% pea protein	25% dehulled peas 20% pea protein		Thiessen *et al.* 2004

TABLE 2.4.3.5-1 Executive summary of studies testing pea ingredients in aquafeeds for various species (*cont.*)

Species	Initial mean weight (g)	Product	Replaced resource	Inclusion rate (%)	Recommendable inclusion rate (%)	Note	Reference s. App. 1
	~37	Pea meal	FM and wheat meal	5, 15 + 25	Not recommended	Affected immunological antiviral response	Hernández *et al.* 2013
	~19	Mix of plant proteins (PP) with extruded peas	FM	50% PP (8% pea), 75% PP (12% pea) + 100% PP (16.3%) pea	50% PP (12% pea)		Alami-Durante *et al.* 2010
European sea bass (*Dicentrarchus labrax*)	~10	Pea seed meal	FM and corn starch	20 + 40	20	At 40% growth was not impaired, but carbohydrate digestibtilty, digestible energy, and carcass composition in terms of lipid deposition was sign. reduced	Gouveia & Davies 1998

TABLE 2.4.3.5-1 Executive summary of studies testing pea ingredients in aquafeeds for various species (*cont.*)

Species	Initial mean weight (g)	Product	Replaced resource	Inclusion rate (%)	Recommendable inclusion rate (%)	Note	Reference s. App. 1
	~6	Extruded dehulled pea seed meal (Aquatex), Sotexpro, Reims France	FM and corn starch	10. 20 + 30	30	Lower ANF content and enhanced nutritional profile compared to native pea meal	Gouveia & Davies 2000
	Exp. I: ~152 Exp. II: ~100	Fine pea (chickpea and faba bean) flour (1.5mm)	Wheat	Exp. I: 15 + 30 Exp. II: 17 + 35	Exp. I: 30 Exp. II: 35	Increase in gastrointestinal evacuation time, slower increase and decrease in serum glucose levels, high level inclusion affects SGR	Adamidou 2008
Gilthead seabream (*Sparus aurata*)	~93	Fine pea (chickpea and faba bean) flour (1.5mm)	Wheat	Exp. III: 17 and 35	35	No improvements	Adamidou 2008

TABLE 2.4.3.5-1 Executive summary of studies testing pea ingredients in aquafeeds for various species (*cont.*)

Species	Initial mean weight (g)	Product	Replaced resource	Inclusion rate (%)	Recommendable inclusion rate (%)	Note	Reference s. App. 1
Black sea bream (*Acanthopagrus schlegeli*)	~13	(Pea (P) and lupine(L)) protein concentrate	FM	L/P ratios: 3:0, 2:1, 1:2, and 0:3; inclusion levels: 30 and 50% protein /kg dietary protein	50	Amino acid supplementation! All inclusion levels recommendable. At higher inclusion, combinations with more PPC are preferred, while high inclusion of LPC resulted in less efficient feed conversion.	Zhang *et al.* 2012a/b
	~8	Pea protein isolate (PPI), pea protein concentrate (PPC)	FM	0, 8, 16	Min. 16		Gao *et al.* 2013
Australian silver perch (*Bidyanus bidyanus*)	~81	Pea (lupin, faba bean, chick pea, vetch, cow pea)	Reference diet	30		Amino acid availability data	Allan *et al.* 2000a

TABLE 2.4.3.5-1 Executive summary of studies testing pea ingredients in aquafeeds for various species (*cont.*)

Species	Initial mean weight (g)	Product	Replaced resource	Inclusion rate (%)	Recommendable inclusion rate (%)	Note	Reference s. App. 1
	Exp. I: ~16.3 Exp. II: ~18.1 Exp. III: ~11.1	Field pea (faba bean, chick pea, vetch) Whole, dehulled, protein concentrates	Reference diet	30	Min. 30%. but dehulling or protein concentrates	Dehulling improves dry matter and energy digestibility, protein concentrates have much higher protein digestibility	Booth *et al.* 2001
Nile tilapia (*Oreochromis niloticus*)	~2	Pea protein isolate	FM	0, 30, 45 + 60	30	Supplementation with limiting essential amino acids or by enzymatic degradation of phytic acid might even improve applicability	Schulz *et al.* 2007

TABLE 2.4.3.5-1 Executive summary of studies testing pea ingredients in aquafeeds for various species (*cont.*)

Species	Initial mean weight (g)	Product	Replaced resource	Inclusion rate (%)	Recommendable inclusion rate (%)	Note	Reference s. App. 1
African catfish (*Clarias gariepinus*)	~52	Thermally treated pea seed meal	Soya		30	Thermal treatments increase weight gain, most effective is moist heat treatment	Davies & Gouveia 2008
Common carp (*Cyprinus carpio*)	~7.5	Native, autoclaved, and dry cooked pea seed meal	Soybean meal	40–46	40% dry heat-treated pea seed meal	Dry heat treatment more effective	Davies & Gouveia 2010
Indian grass carp (*Labeo rohita*)	~5	Leaf meal supplemented with cellulase and xylanase	Rice bran	15 + 30 with and without enzyme supplement	30% with enzyme supplementation	Rice bran is limited in its availability	Nottanalan *et al.* 2021
Pacific white shrimp (*Litopenaeus vannamei*)	Exp. I: ~5 Exp. II: ~0.7 Exp. III: ~1	Whole and dehulled raw flours	Exp. I: reference diet Exp. II: wheat Exp. III: wheat + soya	Exp. I: 30 Exp. II: 25 Exp. III: 5, 10, 20	30	Dehulling, extrusion, and micronization improve protein and energy digestibility	Davis *et al.* 2002

TABLE 2.4.3.5-1 Executive summary of studies testing pea ingredients in aquafeeds for various species (*cont.*)

Species	Initial mean weight (g)	Product	Replaced resource	Inclusion rate (%)	Recommendable inclusion rate (%)	Note	Reference s. App. 1
	~0.6	Meal	FM	10	10	No differences in growth, weight gain, survival or feed conversion ratio	Roy *et al.* 2009
	~0.1	Meal	FM	0–85	34	Good nitrogen turnover from pea, protein deficiency might be solved using concentrates	Martínez-Rocha *et al.* 2013
Tiger shrimp (*Penaeus monodon*)	~0.02	Meal	Defatted soybean meal	0–42	42	Pea meal can replace soy in shrimp diets	Bautista-Teruel *et al.* 2003

TABLE 2.4.3.5-1 Executive summary of studies testing pea ingredients in aquafeeds for various species (*cont.*)

Species	Initial mean weight (g)	Product	Replaced resource	Inclusion rate (%)	Recommendable inclusion rate (%)	Note	Reference s. App. 1
Blue shrimp (*Litopenaeus stylirostris*)	~0.3	Whole, dehulled, extruded, dehulled – extruded and micronized feed peas	Soy and wheat	30	30	No effect of dehulling, micronized pea diet produced the highest feed intake and growth rate	Cruz-Suarez *et al.* 2001

Pea protein concentrate was shown to be a valuable protein ingredient for salmonids (Øverland *et al.*, 2009) and could replace 20% of high-quality fishmeal protein in diets for Atlantic salmon (*Salmo salar*). It also serves as a starch source in salmon feed without any adverse effect on growth performance. Additionally, pea protein concentrate inclusion improved the physical quality of the extruded salmon diets. The upper inclusion limit for pea protein concentrate has been shown to be below 350 g/kg, above which it induces ongoing irritation and damage to the distal intestine (Penn *et al.*, 2011).

Experiments with different pea products fed to rainbow trout (*Oncorhynchus mykiss*) showed that raw whole and also dehulled peas contained highly digestible protein but the starch was mainly indigestible (Ji, 2017). Procedures like air classification of dehulled peas results in a very valuable pea ingredient with higher concentration of protein and a reduced starch content which also resulted in superior digestibility relative to whole, dehulled, or extruded peas for rainbow trout (Thiessen *et al.*, 2003; Thiessen *et al.*, 2004). Pea meal and pea protein concentrate can be included in feed for rainbow trout at levels ≥ 300 g/kg without impairing performance, considering the nutritional value of the ingredient.

Different fish species exhibit distinct ability to digest pea products. The protein digestibility of extruded peas is lower in trout (88%) than in turbot (92%) (Burel *et al.*, 2000). Similarly, the digestibility (apparent digestibility coefficient (ADC)) of energy, mainly supplied as starch, was lower in trout (69%) than in turbot (78%). In addition, extrusion and dehulling have a positive effect on the nutritional value of peas (as well as chickpeas and faba beans), increasing their potential as a feed ingredient, with a higher value for European seabass (*Dicentrarchus labrax*) than for gilthead seabream (*Sparus aurata*), mainly as a carbohydrate substitute for wheat and to a lesser extent as a protein substitute (Gouveia and Davies, 1998; Gouveia and Davies, 2000; Adamidou, 2008).

In feed for the black sea bream (*Acanthopagrus schlegelii*), pea protein concentrates and isolates proved a promising alternative for fishmeal and was recommendable at least at an inclusion level of 16% and can replace 50% of the fishmeal protein (Zhang *et al.*, 2012; Zhang *et al.*, 2012; Gao *et al.*, 2013).

For juvenile milkfish (*Chanos chanos*) pea meal can constitute up to 20% of the total dietary protein in diets (Borlongan *et al.*, 2003).

The Australian silver perch (*Bidyanus bidyanus*) (Allan *et al.*, 2000a; Allan *et al.*, 2000b; Stone *et al.*, 2000; Booth *et al.*, 2001; Stone *et al.*, 2003) is capable of digesting protein effectively but legumes in general tend to have lower ADCs for some amino acids e.g., sulfur-containing amino acids. Individual amino acid availability data as well as exact species requirements allow the balancing of amino acids, facilitating legume inclusion. A combined study about the least-cost formulation of practical diets revealed that a combination of a variety of (protein) resources (fish meal, meat meal, blood meal, field pea, lupin), resulting in the same performance, can be significantly more cost effective than other tested diets. Furthermore, the amino acid profile of the feed generally profits from mixing several ingredients.

For Nile tilapia (*Oreochromis niloticus*), an inclusion of 30% pea protein isolate can be recommended and considering a supplementation with the limiting essential amino acids and/or degradation of phytic acid may further improve the applicability (Schulz *et al.*, 2007). African catfish (*Clarias gariepinus*) and common carp (*Cyprinus carpio*) benefited from dry heat-treated pea seed meal in their diets, compared to untreated or wet heat-treated peas, showed better growth performance, feed utilisation efficiency and apparent digestibility, probably due to thermal reduction of trypsin inhibitory activity (Davies and Gouveia, 2008; Davies and Gouveia, 2010).

Nottanalan *et al.* (2021) incorporated the leafs of peas in diets of Indian Grass Carp (*Labeo rohita*) fingerlings to replace de-oiled rice bran, which is gaining more and more attention as life stock feed and in the human food sector, which forces fish farmers to find cheap alternatives. An inclusion of 300 g/kg pea leafs supplemented with exogenous cellulase and xylanase (1:1) at 1 g/kg showed to be adequate. That means, pea leaf meal can completely replace the rice bran in the diet of *L. rohita* without compromising the growth performance.

Research about the applicability of pea in aquafeed is not restricted to finfish, since several studies also have been conducted with different shrimp species. Various processing methods (dehulling, extrusion, and micronization) have been applied to test the effect on biological availability of nutrients from peas for the Pacific white shrimp (*Litopenaeus vannamei*) (Davis and Arnold, 2000; Davis *et al.*, 2002). The authors draw the conclusion that based on the positive results from their study commercial feed producers should definitely consider the use of peas as an alternative protein and energy source if they are suitably priced. When properly processed (extruded or micronized), protein and energy from peas is highly digestible and there are no adverse effects detectable on shrimp growth and survival, at the inclusion levels tested, which were up to 30% of the diet replacing wheat and some of the soy meal. For shrimp as for fish, the balancing of the single amino acids (sulfur-containing amino acids, threonine, lysine, histidine) is crucial when it comes to the application of plant based proteins, a simple balance of nitrogen content would not be sufficient (Martínez-Rocha *et al.*, 2013). The authors concluded that nutritionally balanced diets formulated with fully complemented amino acid profiles using fishmeal and pea meal derivates (isolated or concentrated protein) at higher protein levels may support higher growth.

Peas proved to be a valuable feed ingredient resource for Tiger shrimp (*Penaeus monodon*) as well. Inclusion rates of peas up to 25% of the total dietary protein, replacing mainly soya in feeds for post larvae (0.02 ± 0.01g) for 90 days showed no adverse effects on growth, feed intake, FCR, survival, body composition or digestibility of dry matter or protein (Bautista-Teruel *et al.*, 2003). For the Blue shrimp

(*Litopenaeus stylirostris*) the whole raw pea was used as feed ingredient and proved to be a very acceptable ingredient at inclusion rates of 30%. Extrusion cooking or micronizing improves the peas nutritional value, while dehulling had no effect (Cruz-Suarez *et al.*, 2001).

In the study by Øverland *et al.* (2009) mentioned earlier in this chapter, an improvement of the physical quality of the extruded salmon diets containing pea protein concentrate could be attained, showing a higher pellet durability index and a higher pellet breaking force than diets with conservative ingredients (soybean meal/fishmeal). Based on these findings, Hansen *et al.* (2011) made feed mixtures to replace fishmeal with deshelled krill meal which is known to reduce the expansion of feed pellets due to its high fat content. Hansen and colleagues tried to adjust the physical pellet properties by inclusion of pea protein and found increased pellet durability and decreased water stability with increasing rates of the krill meal/pea meal mixture.

Peas (as well as chickpeas and faba beans) are raw materials that could be used up to 350 g/kg in seabass and seabream diets without causing negative effects on pellet physical characteristics, producing pellets that could be manufactured, transferred and stored.

Starch, the binder for extruded feed pellets, is mainly supplied to diets in the form of wheat. Pea starch and wheat starch have different processing characteristics, and the extrusion conditions need to be adjusted according to the starch source used when producing extruded fish feed (Sørensen *et al.*, 2011). The physical characteristics of pea meal seems to be similar to that of soya, since increasing replacement of defatted soybean meal with pea meal did not change pellet water stability even at the highest level of replacement (Bautista-Teruel *et al.*, 2003).

References

Adamidou S., 2008: Effect of extrusion on the nutritional value of peas (*Pisum sativum*), chickpeas (*Cicer arietinum*) and faba beans (*Vicia faba*) and inclusion in feeds for European seabass (*Dicentrarchus labrax*) and gilthead seabream (*Sparus aurata*). Doctoral thesis. University of Sterling.

Allan G.L., Parkinson S., Booth M.A., Stone D.A.J., Rowland S.J., Frances J., Warner-Smith R., 2000a: Replacement of fish meal in diets for Australian silver perch, *Bidyanus bidyanus*: I. Digestibility of alternative ingredients. Aquaculture, 186, 293–310.

Allan G.L., Rowland S.J., Mifsud C., Glendenning D., Stone D.A.J., Ford A., 2000b: Replacement of fish meal in diets for Australian silver perch, *Bidyanus bidyanus*: V. Least-cost formulation of practical diets. Aquaculture, 186, 327–340.

Bautista-Teruel M.N., Eusebio P.S., Welsh T.P., 2003: Utilization of feed pea, *Pisum sativum*, meal as a protein source in practical diets for juvenile tiger shrimp, *Penaeus monodon*. Aquaculture, 225, 121–131.

Bellof G., Freitag M., 2021: Körnerleguminosen. Editors: Durst L., Freitag M., Bellof G, 2021: Futtermittel für landwirtschaftliche Nutztiere. DLG Verlag Frankfurt/Main, 258–304.

Bellof G., Halle I., Rodehutscord M., 2020: UFOP-Praxisinformationen: Ackerbohnen, Körnerfuttererbsen, Süßlupinen und Sojabohnen in der Geflügelfütterung. Editor: Union zur Förderung der Öl- und Proteinpflanzen e.V. (UFOP). Berlin. 2nd updated edition.

Bengala-Freire J., Hulin J.C., Peiniau J., Aumaitre A., 1989: Effet de la cuisson extrusion du pois de printemps sur 1a digestibilite des aliments de sevrage precoce du porcelet et consequences sur les performances jusqu'a l'abattage. Journ. Rech. Porcine France, 21, 75–82.

Bertrand G., Perez J.M., Quemere P., 1980: Utilisation du pois proteagineux par le porcelet sevre precocement. Influence de la nature de la cereale. Journ. Rech. Porcine France, 12, 215–226.

Booth M.A., Allan G.L., Frances J., Parkinson S., 2001: Replacement of fish meal in diets for Australian silver perch, *Bidyanus bidyanus*: IV. Effects of dehulling and protein concentration on digestibility of grain legumes. Aquaculture, 196, 67–85.

Borlongan I.G., Eusebio P.S., Welsh T., 2003: Potential of feed pea (*Pisum sativum*) meal as a protein source in practical diets for milkfish (*Chanos chanos* Forsskal). Aquaculture, 225, 89–98.

Burel C., Boujard T., Tulli F., Kaushik S.J., 2000: Digestibility of extruded peas, extruded lupin, and rapeseed meal in rainbow trout (*Oncorhynchus mykiss*) and turbot (*Psetta maxima*). Aquaculture, 188, 285–298.

Ciurescu G., Pană C.O., 2017: Effect of dietary untreated field pea (*Pisum sativum L.*) as substitute for soybean meal and enzymes supplementation on egg production and quality of laying hens. Romanian biotechnological letters, 22, 12204–12213.

Coenen M., 2020: Futtermittel. Editors: Coenen M. and Vervuert I., 2020: Pferdefütterung. Georg Thieme Publisher. Stuttgart – New York. 6th revised edition.

Cruz-Suarez L.E., Ricque-Marie D., Tapia-Salazar M., Mccallum I.M., Hickling D., 2001: Assessment of differently processed feed pea (*Pisum sativum*) meals and canola meal (*Brassica* sp.) in diets for blue shrimp (*Litopenaeus stylirostris*). Aquaculture, 196, 87–104.

Czarnecki R., Jacyno G., Lubowicki R., Petkov K., Delecta A., 1988: Influence of using pea in feeding young boars on their growth rate, backfat thickness and quantity and quality of semen. In: Proc. VIth world conf. Anim. Prod., Helsinki: 604.

Dahl W.J., Foster L.M., Tyler R.T., 2012: Review of the health benefits of peas (*Pisum sativum* L.). British journal of nutrition, 108, 3–10.

Davies S., Gouveia A., 2008: Enhancing the nutritional value of pea seed meals (*Pisum sativum*) by thermal treatment or specific isogenic selection with comparison to soybean meal for African catfish, *Clarias gariepinus*. Aquaculture, 283, 116–122.

Davies S., Gouveia A., 2010: Response of common carp fry fed diets containing a pea seed meal (*Pisum sativum*) subjected to different thermal processing methods. Aquaculture, 305, 117–123.

Davis D., Arnold C., Mccallum I., 2002: Nutritional value of feed peas (*Pisum sativum*) in practical diet formulations for *Litopenaeus vannamei*. Aquaculture nutrition, 8, 87–94.

Davis D.A., Arnold C.R., 2000: Replacement of fish meal in practical diets for the Pacific white shrimp, *Litopenaeus vannamei*. Aquaculture, 185, 291–298.

Ebert U., Berk A., 2013: Buntblühende Wintererbsen in der Schweinefütterung unter den Bedingungen des Ökologischen Landbaus. [Coloured blooming winter peas in pig feed under the conditions of organic farming.] Kompetenzzentrum Ökolandbau Niedersachsen GmbH, D-Visselhövede und Friedrich-Löffler-Institut, D-Braunschweig. https://www.orgprints.org/id/eprint/25293/.

Farrell D.J., Perez-Maldonado R.A., Mannion P.F., 1999: Optimum inclusion of field peas, faba beans, chick peas and sweet lupins in poultry diets. II. Broiler experiments. British poultry science, 40, 674–680.

Gao S., Yin N., Zhou F., Li H., Zhou J., Wang R., Shao Q., 2013: Evaluation of pea proteins and poultry protein as fish meal alternatives in the diets for juvenile black sea bream, *Acanthopagrus schlegelii*. Aquaculture nutrition, 19, 278–288.

Gatel F., Grosjean F., 1990: Composition and nutritive value of peas for pigs: a review of European results. Livestock production science, 26, 155–175.

Gatel F., Grosjean F., Leuillet M., 1988: Utilization of white-flowered smooth-seeded spring peas (*Pisum sativum hortense*, CV Amino) by the breeding sow. Animal feed science and technology, 22, 91–104.

Gatel F., Fekete J., Grosjean F., 1989: A note on the use of spring pea (*Pisum sativum hortense*) in diets for weaned pigs. Anim. Prod., 49, 330–332.

Gdala J., Buraczewska L., 1997: Chemical composition and carbohydrate content of several varieties of faba bean and pea seeds. J. Anim. Feed sci., 6, 123–135.

Gouveia A., Davies S., 1998: Preliminary nutritional evaluation of pea seed meal (*Pisum sativum*) for juvenile European sea bass (*Dicentrarchus labrax*). Aquaculture, 166, 311–320.

Gouveia A., Davies S., 2000: Inclusion of an extruded dehulled pea seed meal in diets for juvenile European sea bass (*Dicentrarchus labrax*). Aquaculture, 182, 183–193.

Hansen J.Ø., Shearer K.D., Øverland M., Penn M.H., Krogdahl Å., Mydland L.T., Storebakken T., 2011: Replacement of LT fish meal with a mixture of partially deshelled krill meal and pea protein concentrates in diets for Atlantic salmon (*Salmo salar*). Aquaculture, 315, 275–282.

Hoffmann M., Steinhöfel O., 2023: Futtermittel- und rationsspezifische Restriktionen. Agrimedia. Erling Verlag GmbH & Co. KG.

Janocha A., Milczarek A. Głuchowska J. 2022: Efficiency of peas in broiler chicken feeding. Animal Science and Genetics, 18, 11–23.

Janusonis S., Benediktaviciute-Kiskine A., Juodka R., 2009: Efficiency of peas in the diets of heavy-type (BIG-6) turkeys. Animal husbandry, 54, 100–106.

Ji B., 2017: Determination of the effects of pulse starches (pea, lentil and faba bean) versus corn starch on growth performance and glucose tolerance in rainbow trout (*Oncorhynchus mykiss*). Master thesis. University of Saskatchewan.

Juodka R., Nainiene R., Juskiene V., Juska R., Stuoge I., 2016: Effects of different amounts of field peas (*Pissum sativum L.*) in the diets for turkeys on meat qualities. Journal of applied animal research, 44, 150–157.

Kirn A.I., Weindl P., Bellof G., Vahjen W., Zentek J., Weber M., 2022: Erbsen und Erbsenprodukte in der Hühnermast und Ferkelaufzucht unter besonderer Berücksichtigung der Darmgesundheit. Union zur Förderung der Oel- und Proteinpflanzen e.V. (UFOP). Final report of the project 524/211.

Kuhnitzsch C., Steinhöfel O., 2021: Erbsen für Kühe? top agrar 1: R18–21.

Le Guen M.P., Huisman J., Gueguen J., Beelen G., Verstegen M.W.A., 1995: Effects of a concentrate of pea antinutritional factors on pea protein digestibility in piglets. Livestock production science, 44, 157–167.

Martínez-Rocha L., Gamboa-Delgado J., Nieto-López M., Ricque-Marie D., Cruz-Suárez L.E., 2013: Incorporation of dietary nitrogen from fish meal and pea meal (*Pisum sativum*) in muscle tissue of Pacific white shrimp (*Litopenaeus vannamei*) fed low protein compound diets. Aquaculture research, 44, 847–859.

Nikolopoulou D., Grigorakis K., Stasini M., Alexis M.N., Iliadis K., 2007: Differences in chemical composition of field pea (*Pisum sativum*) cultivars: effects of cultivation area and year. Food Chemistry, 103, 847–852.

Nottanalan H., Kumar S., Sardar P., Varghese T., Maiti M.K., Srivastava P.P., Sahu N.P., 2021: Green pea, *Pisum sativum* leaf meal with supplemented exogenous cellulase and xylanase can completely replace de-oiled rice bran (DORB) in the diet of *Labeo rohita* (Hamilton, 1878) fingerlings. Aquaculture research, 52, 3186–3197.

Offner A., Bach A., Sauvant D., 2003: Quantitative review of in situ starch degradation in the rumen. Anim. Feed sci. Technol., 106, 81–93.

Øverland M., Sørensen M., Storebakken T., Penn M., Krogdahl Å., Skrede A., 2009: Pea protein concentrate substituting fish meal or soybean meal in diets for Atlantic salmon (*Salmo salar*) – effect on growth performance, nutrient digestibility, carcass composition, gut health, and physical feed quality. Aquaculture, 288, 305–311.

Owusu-Ansah Y.J., Mccurdy S.M., 1991: Pea proteins: a review of chemistry, technology of production, and utilization. Food reviews international, 7, 103–134.

Penn M.H., Bendiksen E.Å., Campbell P., Krogdahl Å., 2011: High level of dietary pea protein concentrate induces enteropathy in Atlantic salmon (*Salmo salar* L.). Aquaculture, 310, 267–273.

Preissinger W., Obermeier A., Spiekers H., 2007: Zum Einsatz hydrothermisch behandelter Erbsen in der Milchviehfütterung. UFOP-Schriften 33, 35–43.

Roy L.A., Bordinhon A., Sookying D., Davis D.A., Brown T.W., Whitis G.N., 2009: Demonstration of alternative feeds for the Pacific white shrimp, *Litopenaeus vannamei*, reared in low salinity waters of west Alabama. Aquaculture Research 40(4), 496–503.

Savage G.P., Deo S., 1989: The nutritional value of peas (*Pisum sativum*): a literature review. Nutrition abstracts and reviews (series A), 59, 65–88.

Schulz C., Wickert M., Kijora C., Ogunji J., Rennert B., 2007: Evaluation of pea protein isolate as alternative protein source in diets for juvenile tilapia (*Oreochromis niloticus*). Aquaculture Research, 38, 537–545.

Şentürklü S., Landblom D.G., 2015: Effect of field pea replacement for oats on palatability, feeding safety, and growth performance in yearling American quarter horses. Kafkas Univ. Vet. Fak. Derg, 21, 447–454.

Sørensen M., Morken T., Kosanovic M., Øverland M., 2011: Pea and wheat starch possess different processing characteristics and affect physical quality and viscosity of extruded feed for Atlantic salmon. Aquaculture nutrition, 17, e326–e336.

Steinhöfel O., Kuhnitzsch C., 2021: Untersuchungen zur Verbesserung ernährungsphysiologischer Eigenschaften von Ackerbohnen und Erbsen für Milchrinder durch kombinierte gärbiologische und thermische Behandlungen (SilaToast). Final report of the BLE project FKZ: 2815EPS020.

Stone D.A.J., Allan G.L., Parkinson S., Frances J., 2003: Replacement of fishmeal in diets for Australian silver perch *Bidyanus bidyanus* (Mitchell). II. Effects of cooking on digestibility of a practical diet containing different starch products. Aquaculture research, 34, 195–204.

Stone D.A.J., Allan G.L., Parkinson S., Rowland S.J., 2000: Replacement of fish meal in diets for Australian silver perch. *Bidyanus bidyanus*: III. Digestibility and growth using meat meal products. Aquaculture, 186, 311–326.

Thiessen D., Maenz D., Newkirk R., Classen H., Drew M., 2004: Replacement of fishmeal by canola protein concentrate in diets fed to rainbow trout (*Oncorhynchus mykiss*). Aquaculture nutrition, 10, 379–388.

Thiessen D.L., Campbell G.L., Adelizi P.D., 2003: Digestibility and growth performance of juvenile rainbow trout (*Oncorhynchus mykiss*) fed with pea and canola products. Aquaculture nutrition, 9, 67–75.

Thingnes S.L., Gaustad A.H., Kjos N.P., Hetland H., Framstad T., 2013: Pea starch meal as a substitute for cereal grain in diets for lactating sows: the effect on sow and litter performance. Livestock science, 157, 210–217.

Wang T.L., Domoney C., Hedley C.L., Casey R., Grusak M.A., 2003: Can we improve the nutritional quality of legume seeds? Plant physiol., 131, 88–891.

Weber M., Preissinger W., Bellof G., 2020: Praxisinformation "Ackerbohnen, Körnerfuttererbsen, Süßlupinen und Sojabohnen in der Schweinefütterung". Editor: Union zur Förderung von Öl- und Proteinpflanzen e.V. (UFOP). Berlin. 2nd updated editon.

Zeyner A., Bachmann M., Okon P., 2021: Untersuchungen zur Verbesserung ernährungsphysiologischer Eigenschaften von Ackerbohnen und Erbsen für Milchrinder durch kombinierte gärbiologische und thermische Behandlungen. Final report of the BLE project FKZ: 2815EPS058.

Zhang Y., Øverland M., Sørensen M., Penn M., Mydland L.T., Shearer K.D., Storebakken T., 2012a: Optimal inclusion of lupin and pea protein concentrates in extruded diets for rainbow trout (*Oncorhynchus mykiss*). Aquaculture, 344–349, 100–113.

Zhang Y., Øverland M., Xie S., Dong Z., Lv Z., Xu J., Storebakken T., 2012b: Mixtures of lupin and pea protein concentrates can efficiently replace high-quality fish meal in extruded diets for juvenile black sea bream (*Acanthopagrus schlegelii*). Aquaculture, 354–355, 68–74.

2.5 Faba beans (*Vicia faba L.*)

Mechthild Freitag

Faba beans, also known as field beans, horse beans or broad beans, have a long history of cultivation. They were domesticated in the 11th millennium cal BP in the lower Galilee, Israel (Caracuta *et al.*, 2015). These legumes serve both human and animal needs and are major crops in many countries. Globally, faba beans rank as the third most important feed grain legume after soybeans (*Glycine max L.*) and peas (*Pisum sativum L.*). In particular, faba beans are highly efficient N-fixing legumes, contributing to the reduction of N_2O greenhouse gas emissions.

2.5.1 *Ingredients*

2.5.1.1 Nutrients and antinutritive Factors (ANF)

2.5.1.1.1 *Nutrients*

There are two main types of faba beans with significant importance for feeding purposes. They can be distinguished by the colour of their flowers and their seed hulls. White-flowered cultivars have beige or grey seed hulls of varying shades. Colour-flowered cultivars have seed hulls of various colours. However, as the seeds age, the colour of the coat changes and the shades overlap. Colour-flowered varieties tend to have a more pronounced bitter taste due to a higher tannin content. This may limit their inclusion in diets. However, due to their higher yields under adverse growing conditions, they play an important role in organic farming.

Differences in nutrient content primarily arise within the colour-flowered cultivars rather than between white and coloured varieties. Faba beans are a source of both, protein and energy, containing on average about 30% protein in their dry matter – more than twice the protein content of cereals. The protein content varies from about 20 to 40%, depending to the cultivar, soil characteristics, fertilization and weather conditions (moisture, temperature) during growth. Additionally, faba beans have a substantial starch content, comparable to that of hulled oats (table 2.5.1.1-1). Protein and starch content vary considerably between cultivars and are genetically negatively correlated. The starch content must be taken into account in ration design, especially for ruminant rations. Faba bean starch is partly resistant to enzymatic hydrolysis, as evidenced by its significant amount of resistant starch compared to cereals (Bello-Pérez *et al.*, 2007). The sugar content is higher than that of most cereals but lower than that of soybeans (up to 45 g/kg DM). The lipid concentration is also relatively low. The fibre content (NDF, ADF) is similar to that of cereals, ranging from 15 to 30%, with hemicellulose, cellulose and lignin being the main components (Dhull *et al.*, 2021). The fibres are largely available for rumen fermentation. They are fermented to acetic acid, which is the basis for the formation of milk fat.

FIGURE 2.5-1 Faba beans

FIGURE 2.5-2
Colour-flowered faba bean "Trumpet"

TABLE 2.5.1.1-1 Nutrients (g/kg) of white- and colour-flowered faba beans (*Vicia faba L.*)

Nutrient	Faba beans, white-flowered			Faba beans, colour-flowered	
DM	880	1000	1000 (Min–max)	880	1000
CP	274	311	261–330	263	299
CL	11	13	11–23	14	16
CF	77	87	57–115	82	93
$aNDF_{om}$	140	159	114–227	129[1]	147[1]
ADF_{om}	93	106	85–148	110[1]	125[1]
STA	381	433	375–489	390	443
SUG	38	43	11–45	28	32
NFE	470	535	–	489	556
ASH	37	42	32–48	32	37

1 NDF, ADF.
BELLOF AND FREITAG, 2021

FIGURE 2.5.1.1-1
Faba bean seeds
PHOTO: ©CH. ZIRON

Variation in the nutrient content, especially in the protein content are not only due to differences in cultivars but more important to environmental conditions during growth: temperature, humidity, fertilization and soil composition.

2.5.1.1.2 *Carbohydrates*

Even though faba beans are considered a protein feed in animal nutrition, the main components are carbohydrates (table 2.5.1.1-2). Besides starch – the main nutrient – different fibre fractions play an important role.

TABLE 2.5.1.1-2 Carbohydrate content (g/kg DM) in faba beans (*Vicia faba L.*)

Ingredient	Mean	Minimum	Maximum
Total carbohydrates	565	457	701
STA	360	300	417
SUG	43	40	46
NFE	488	381	619
NDF	221	119	426
ADF	116	103	135
CF	97	15	224
ADL	25	19	40

MENG *ET AL.*, 2021, MODIFIED

TABLE 2.5.1.1-3 Ruminal starch degradation in faba beans (*Vicia faba L.*) compared to wheat starch

Treatment	Immediately degradable (%)	Degradability rate (%)
Beans untreated	37.1	75.0
Beans toasted	36.9	64.4
Wheat untreated	60.4	94.2
Wheat byproduct	78.2	95.1

OFFNER *ET AL.*, 2003, MODIFIED

The sugars are mainly soluble and consist in part of oligosaccharides (table 2.5.1.1-4). Total starch has a range of 300–489 g per kg DM. It mainly consists of digestible (enzyme susceptible) and rumen fermented starch. However, the degradability of this starch is much slower than that of wheat and the amount of rumen bypass starch is higher (Offner *et al.*, 2003; table 2.5.1.1-3). There are no significant differences between white- and colour-flowered varieties in this respect but starch and fibre content vary in relation to cultivars and growing conditions. The total amount of fibre does also not differ between the different colour-flowered varieties.

2.5.1.1.3 *Antinutritional factors (ANF)*

For more extensive descriptions of ANF in legumes refer to section 2.3.

Faba beans contain various ANFs whose presence may limit their use in animal diets. They include condensed tannins, phytates, vicine, convicine, saponins, lectins, various oligosaccharides, and trypsin and protease inhibitors. ANF concentrations in raw faba bean seeds have a wide range (table 2.5.1.1-5).

TABLE 2.5.1.1-4 Content and chemical composition of oligosaccharides and non-starch polysaccharides (NSP) in faba beans (*Vicia faba L.*) (g/kg DM)

Item	Faba bean cultivar			
	Alen	**Kamir**	**Martin**	**Tibo**
Oligosaccharides				
Sucrose	15.1	14.1	15.1	16.7
Galactopinitol	1.5	1.4	1.4	1.8
Galactoinositol	1.0	0.9	0.9	1.0
Raffinose	2.1	1.8	2.3	2.2
Stachyose	6.6	5.9	7.3	6.8
Verbascose	15.8	13.8	17.8	18.1
Total a-galactosides	26.9	23.8	29.8	29.8
Total oligosaccharides	42.0	37.9	44.9	46.6
NSP				
Rhamnose	2.3	1.3	1.5	1.9
Fructose	0.9	0.7	0.9	0.9
Arabinose	23.8	23.3	24.3	22.3
Xylose	12.3	13.8	15.3	14.8
Mannose	6.2	2.7	2.1	3.1
Galactose	12.8	10.8	10.6	10.3
Glucose	92.6	95.9	94.4	102.1
Uronic acids	25.8	24.8	25.4	25.6
Total NSP	176.8	173.3	174.5	181.2

GDALA AND BURACZEWSKA, 1997

The main ANFs limiting the use of faba beans in animal feed are condensed tannins and the glucopyranosides vicine and convicine (V/C). Trypsin inhibiting activity (TIA) is generally very low and does not significantly impact the use of faba beans in animal diets. Tannin and V/C content mainly depend on the genotype of the beans and on growing conditions.

The tannin content of faba beans is quite variable and generally more related to flower colour than to seed colour. White-flowered faba beans typically contain negligible amounts (0.1 g/kg DM), whereas colour-flowered beans can have higher levels up to 5–10 g/kg DM (table 2.5.1.1-6; Crépon *et al.*, 2010). Tannins can reduce the apparent digestibility of protein and lysine and increase endogenous protein excretion. They also have a negative impact on the apparent digestibility of starch, resulting in lower energy availability. Dehulling the beans can significantly

TABLE 2.5.1.1-5 Antinutritive compounds (g/kg) of faba bean (*Vicia faba L.*) seed

Ingredient	Mean	Minimum	Maximum
Total phenolics	17.59	10.7	28.48
Tannins	7.38	0.06	14.5
Vicine	5.59	0.99	12.38
Convicine	2.72	1.35	3.88
Phytic acid	9.24	1.7	21.25
Poliphenols	2.49	*	*
Trypsin inhibitor activity	0.9	0.6	1.5
Verbascose	25.79	16.9	36.44
Stachyose	13.44	7.6	18.7
Raffinose	3.26	2	4.5

* No data.

MENG *ET AL.*, 2021

TABLE 2.5.1.1-6 ANF concentration in faba beans (*Vicia faba L.*) with low and high tannin content

Antinutritive component	Tannin content	
	Low	High
Condensed tannins (g/kg)	0.1	6.6
TIA (UTI/mg)	2.9	2.9
Vicine + convicine (g/kg)	7.6	8.3

CRÉPON *ET AL.*, 2010

reduce tannin concentrations, as tannins are mainly concentrated in the outer hull (Föste, 2021).

Besides tannins, vicine and convicine are the main ANFs in poultry diets. Due to erythrocyte haemolysis they reduce egg size. Moreover, V/C negatively affect nutrient digestibility in poultry. Their impact on protein, starch and energy digestibility is coupled with that of tannins. On average, the V/C content is higher in white-flowered than in colour-flowered cultivars.

2.5.1.2 Minerals and vitamins

The mineral content of faba beans is similar to that of cereals. They are rich in phosphorus (P) (1.9–6.0 g/kg DM) and intermediate in calcium (Ca) (1.2–4.3 g/kg DM).

Tannins, especially in colour-flowered cultivars, may restrict mineral digestibility. In general, due to the phytate content, P digestibility in faba beans is less than 40% for pigs. Therefore, the addition of phytases to pig and poultry diets is recommended.

Compared to wheat, faba beans contain approximately twice as much potassium (K) (11 g/kg DM) and iron (Fe) (70–85 g/kg DM). However, there are great variations between cultivars. Sodium (Na) content is low (0.1–2.1 g/kg DM), as is selenium (Se) content (0,02 g/kg DM). Mineral availability depends in part on the phytate concentration of the specific cultivar. Intestinal mineral digestibility of Ca, P, Zn and Fe can be enhanced by adding phytases or by fermentation of beans. The content of trace minerals, e.g. Co, Cu, Fe, J, Mn, Mo, Zn and Se varies considerably, depending on their soil content and biological availability.

Analytic results of faba beans depend on the cultivar, maturity stage, cultivation environment, soil qualities, and even to a greater extent to weather conditions during growth. Therefore, before adding high amounts in diets an analysis of their main nutrients and minerals should be conducted.

Faba beans are a good source of vitamin A as well as of various vitamins B (table 2.5.1.2-2). The vitamin E content is rather low.

2.5.2 *Feed value*

2.5.2.1 Protein

The protein quality for ruminants is determined by the N content, the amount of rumen undegraded protein (RUP) and the amount of essential amino acids in the bypass protein. German feed tables indicate that faba beans have a low percentage of rumen undegraded protein (15% RUP) according to LFL Bayern (2017), while CVB (2018) declares an amount of 20%. Most N compounds are rapidly fermented in the rumen. *In vitro* studies by Bissinger *et al.* (2007) indicated higher amounts of RUP in a tannin-rich variety compared to a cultivar with a low tannin content. However, the rumen passage rate is more important than the tannin content in determining the level of protein degradation in the rumen. Faster passage rates result in higher RUP levels.

As with other nutrients, amino acid composition of faba beans varies greatly and depends on the cultivar and on growing conditions. Faba bean seeds have higher quantities of lysine and arginine, which makes them suitable for combining with cereals to provide essential amino acids (AA), resulting in a more balanced and desirable AA profile of the diet. In total, essential AA account for more than 50% of all AA (Meng *et al.*, 2021). However, methionine, cysteine and tryptophan levels are low (table 2.5.2.1-1).

For monogastric animals, faba beans are also a source of both, protein and energy. The essential AA digestibility for white-flowered faba beans is approximately 83% in pigs and poultry. For colour-flowered cultivars, the protein digestibility is reduced to 69% in pigs and 78% in poultry (table 2.5.2.1-2). In pigs, especially the AAs methionine, cysteine and tryptophan have a low digestibility of 62, 55 and

TABLE 2.5.1.2-1 Average mineral content (in DM) of faba beans (*Vicia faba L.*)

Item		Faba beans, white-flowered	Faba beans, colour-flowered
Ca	g	2.8	1.6
P	g	4.0	5.3
DCP_{Pig}[1]	g	1.5	2.0
$DCP_{Poultry}$	g	*	*
Mg	g	2.1	1.8
Na	g	0.1	0.1
K	g	11	11
Cl	g	1.0	0.7
S	g	2.0	2.8
Co	mg	0.03	0.4
Cu	mg	13	14
Fe	mg	85	68
J	mg	0.2	*
Mn	mg	8.1	8.1
Mo	mg	0.7	0.7
Zn	mg	37	36
Se	mg	0.02	*
DCAB[2]	meq	172	102
BE[3]	mmol/kg	240	272

* No data.
1 Digestible P.
2 Dietary cation-anion balance.
3 Base excess.
BELLOF AND FREITAG, 2021

TABLE 2.5.1.2-2 Vitamin content of faba beans (*Vicia faba L.*)

Vitamin		White-flowered	Colour-flowered
Vitamin A	mg/kg	589	50
Vitamin E	mg/kg	5.8	5.8
Thiamin (B_1)	mg/kg	7.4	5.8
Riboflavin (B_2)	mg/kg	3.7	3.5
Niacin	mg/kg	32	29
B_6	mg/kg	4.1	4.3
B_{12}	mg/kg	0	0

TABLE 2.5.1.2-2 Vitamin content of faba beans (*Vicia faba L.*) (*cont.*)

Vitamin		White-flowered	Colour-flowered
Pantothenic acid	mg/kg	4.0	3.5
Cholin	mg/kg	192	192
Biotin	µg/kg	105	104
Folic acid	mg/kg	4.7	*

* No data.

BELLOF AND FREITAG, 2021

TABLE 2.5.2.1-1 Amino acid (AA) content of white- and colour-flowered faba beans (*Vicia faba L.*)

Amino acid	White-flowered			Colour-flowered	
	g/kg (88% DM, 274 g CP)	g/100g CP	g/100g CP (Min–max)	g/kg (88% DM, 263 g CP)	g/100g CP
Lys	17.6	6.4	5.9–6.4	16.4	6.2
Met	1.9	0.7	*	1.7	0.7
Cys	3.5	1.3	*	3.1	1.2
Thr	9.7	3.5	3.5–3.7	9.2	3.5
Trp	2.2	0.8	0.8–1.1	2.3	0.9
Ile	11.1	4.1	*	10.4	4.0
Leu	20.7	7.6	*	18.9	7.2
Val	12.4	4.5	*	11.6	4.4

* No data.

BELLOF AND FREITAG, 2021

57%, respectively. This exacerbates the problem of low AA levels. Amino acids need to be balanced in the diet or reduced growth rates will be observed.

The levels of essential AAs of white- and colour-flowered faba beans are listed in table 2.5.2.1-1 and their digestibility for pigs and poultry in table 2.5.2.1-2. Both, AA content and their digestibility, vary according to the cultivar and growing conditions. Digestibility also depends on the age of animals.

2.5.2.2 Energy

In spite of a high crude fibre content, the energy digestibility is high. This can be explained by the high digestibility of the nitrogen-free extract, crude fibre and

TABLE 2.5.2.1-2 Amino acid digestibility (%) for pigs and poultry of white- and colour-flowered faba beans (*Vicia faba L.*)

Amino acid	White-flowered		Colour-flowered	
	Pig	Poultry	Pig	Poultry
Lys	88	91	81	86
Met	83	83	62	87
Cys	76	74	55	72
Thr	83	87	71	78
Trp	81	71	57	66
Ile	85	88	77	80
Leu	86	91	77	80
Val	82	87	74	79

BELLOF AND FREITAG, 2021

crude protein. Despite the different tannin concentrations, the digestible energy content of colour-flowered faba beans appears to be the same as for white-flowered varieties. In ruminant nutrition, it is comparable to wheat. The relative energy value for pigs and poultry is somewhat lower and compares to hulled barley.

2.5.3 *Feeding recommendations*

Legumes in general are a valuable protein source and feed material for most domestic animals. However, due to their antinutritive factors, acceptable amounts in feed composition vary, depending on the legume species and variety as well as on animal species and age. Over the past few decades, several feeding trials have been carried out to determine the acceptable level of faba beans in farm animal diets.

However, when using faba beans in compound feed, it must be kept in mind that their seeds are larger and therefore more difficult to dry than cereal grains. Residual moisture above 14%, for example due to overgrowth or uneven ripening, can cause mould growth. Animals reject these faba beans. Besides other ANFs, mould can be a reason for reduced feed intake in diets with faba beans.

2.5.3.1 Cattle and small ruminants

To assess the feed value of plants for ruminants, their influence on rumination activity and their effects on rumen fermentation must be taken into account, in addition to the supply of nutrients. The stimulation of chewing and rumination depends on the particle size on the one hand and, on the other hand, on the mechanical resistance to chewing. The latter is estimated by the content of structurally active fibres (NDF, ADF, CF) and their chemical composition (table 2.5.3.1-1). The fibre content

TABLE 2.5.3.1-1 Protein parameters of two different faba bean (*Vicia faba L.*) cultivars (*in vitro* analysis)

Parameter		**Faba bean cultivar**			
		Samba (high tannin)		**Valeria (low tannin)**	
Intake of faba beans		30% of concentrate DM			
Estimated rumen passage rate	%/h	8	5	8	5
CP	g/kg DM	309		339	
RUP	g/kg DM	74	40	60	29
Microbial protein	g/kg DM	134	134	121	111
Ruminal degradation	%	68.0	74.9	72.7	80.4
nXP	g/kg DM	134	134	121	111

BISSINGER *ET AL.*, 2007

of faba beans is much higher than of cereals, especially the ADF content (approximately 159 g $aNDF_{om}$ and 106 g ADF_{om} per kg DM in white-flowered varieties). It can be compared to that of soybeans of European origin. However, as faba beans are milled for feeding purposes, their fibre does not enhance rumination. On the other hand, a larger part of the fibres consists of easily fermentable components (table 2.5.3.1-2) and serves as an energy source for microorganisms in the rumen.

In ruminant feeding, protein quality is mainly determined by the nitrogen content and by the speed and extent of ruminal protein degradation. The AA composition is only considered relevant in the rumen bypass protein. The nitrogen content corresponds to the crude protein value (CP/6.25).

The rate of ruminal fermentation depends on the level of feed intake, which determines the length of time the feed remains in the rumen. It is also influenced by the protein structure and its incrustation with poorly fermentable fibres. A certain amount of protein passes through the rumen unchanged und is digested in the small intestine. This is reflected in the RUP value. Its AA composition has a direct impact on the blood AA levels. For high yielding cows, a certain level of RUP in the diet is essential. In white-flowered faba beans, RUP is low, even lower than in wheat (15%). Colour-flowered faba beans with a higher tannin content have about the same RUP as wheat (20%). However, RUP enhancement of the diet by high tannin cultivars had no positive effect on milk yield or milk protein content (Bissinger *et al.*, 2007).

A feeding trial with both bean cultivars as sole protein source revealed significant differences in rumen fermentation. Feeding the variety with the lower tannin content resulted in a significantly lower pH because of enhanced volatile fatty acid

TABLE 2.5.3.1-2 Rumen fermentation characteristics and milk composition after feeding faba beans (*Vicia faba*) with low or high tannin content

Parameter		Faba bean cultivar	
		Samba (high tannin)	Valeria (low tannin)
pH		6.7[a]	6.4[b]
VFA[1]	mmol/l	90.7[a]	97.0[b]
Acetic acid	%	62.5[a]	57.9[b]
Propionic acid	%	17.8[a]	21.8[b]
NH_3	mmol/l	12.4	11.8
Feed intake / day	kg DM	21.5	21.3
Milk yield / day	kg	27.6	27.6
Milkfat	%	4.33	4.30
Milkprotein	%	3.64[a]	3.75[b]
FECM / d[2]	kg	29.4	29.5
Urea	mg/l	297[a]	318[b]

1 Volatile fatty acids.
2 Fat and energy corrected milk yield.
a,b $p < 0.05$.
BISSINGER *ET AL.*, 2007

and especially propionic acid content (table 2.5.3.1-2). Protein turnover, analysed by NH_3 concentration, did not differ. Milk yield and milk fat content were not affected but – as expected because of the increased level of propionic acid – milk protein concentration was significantly higher when feeding the diet with the lower tannin content.

Johnston *et al.* (2019) fed faba beans (4.1 and 8.4 kg per day) of a high tannin cultivar (Fuego), substituting other protein and starch sources. Even with the amount of 8.4 kg faba beans/cow/day no negative effects on feed intake, milk yield, body tissue reserves or fertility were observed. However, milk fat and milk protein content were both significantly reduced, even though there was no difference in diet digestibility. The authors attributed the effect on the milk fat production to changes in rumen fatty acid production due to differences in rumen starch degradability between cereals and faba beans. They attributed the differences in protein production to a lack of lysine and methionine. The inclusion of high levels of faba beans obviously requires a balance of RUP amino acids in the feed ration. The authors concluded that there are no limits in including faba beans in the diet for high yielding dairy cows if the diet contains enough fermentable energy and the metabolic amino acid profile is well balanced. Faba beans can serve as the sole protein source in ruminant feeding.

TABLE 2.5.3.1-3 Recommendations for the inclusion of faba beans (*Vicia faba L.*) in cattle feed

Animal group	g/100 kg BW[1]	kg/animal	% of compound feed
Calf (up to 4 months of age)	*	*	20
Heifer, from 4 months on			
White-flowered	300	*	25
Colour-flowered	200	*	20
Suckler cow			
White-flowered	600	2.0 (300 kg BW)	*
		3.0 (500 kg BW)	*
		4.0 (700 kg BW)	*
Colour-flowered	500	1.5 (300 kg BW)	*
		2.5 (500 kg BW)	*
		3.5 (700 kg BW)	*
Beef cattle, either variety	300	1.5 (450 kg BW)	30

1 BW: body weight.
* No recommendation.
HOFFMANN AND STEINHÖFEL, 2023

Recommendations of inclusion of faba beans into feed for young cattle amount to 25% and 20% of white- and colour-flowed varieties, respectively, in concentrates, but older cattle will accept both cultivars at a level of 30%. Suckler cows can be fed up to 0.6 kg/100 kg of body weight per day (Hoffmann and Steinhöfel, 2023; table 2.5.3.1-3).

Recommendations for beef cattle are 2.0 kg per day for young animals, but this can be increased to 2.5 kg per day in combination with maize silage. Because of their high starch content, faba beans can substitute soybeans or rapeseed and cereals in compound feed for beef cattle with average daily weight gain of less than 1200 g. Fast growing beef cattle with daily weight gain above 1200 g need more bypass protein. Therefore, a combination of beans with soybeans or rapeseed is recommended.

Young sheep and goat lambs accept up to 20% of white-flowered and 15% of colour-flowered faba beans in compound feed (table 2.5.3.1-4). For mature sheep and dairy goats the concentration of faba beans in the compound feed depends on the protein content of the roughage and the daily protein requirement.

In a feeding trial with sheep fattening, soybean meal and barley were partly substituted by faba beans (200 g faba beans/kg feed, DM) without balancing the essential amino acid levels. The resulting diets were similar in nutrients except for a higher fat content in the diet with faba beans (18.9 versus 23.9 g/kg DM). Dry matter intake and thus nutrient intake was higher in the faba bean group which resulted

TABLE 2.5.3.1-4 Recommendations for the inclusion of faba beans in feed for small ruminants

Animal group	Variety	g/100 kg BW[1]	g/animal/day	% of feed
Sheep				
	White-flowered	360	250 (70 kg BW)	*
	Colour-flowered	290	200	*
Lambs (sheep and goat)				
	White-flowered	*	*	20 (concentrates)
	Colour-flowered	*	*	15 (concentrates)
Dairy goats				
	White-flowered	335	200	
	Colour-flowered	250	150	

1 BW: body weight.
* No recommendation.
HOFFMANN AND STEINHÖFEL, 2023

in a significantly higher daily weight gain (195.8 versus 229.9 g; p = 0.028). Feed conversion ratio did not differ (Al Shabuol and Obeidat, 2022).

2.5.3.2 Pigs

In pig nutrition, feed quality is mainly determined by the energy content and the level of digestible protein and digestible essential amino acids. These are predominantly lysine, methionine, threonine, tryptophan and valine. In total, the sum of essential AAs in faba beans is only half that of soya. Compared on the same protein level, AA concentrations do not differ much to soybean meal except for methionine and tryptophan. Nevertheless, diets have to be balanced for essential AA. The level of each AA is highly correlated to the protein content independent of genotype or environment.

In faba beans, the main ANF is tannin. Levels of trypsin inhibitors, lectins and α-galactosides are low so that a heat treatment is not required. At a level of 200 g faba beans/kg feed, a tannin content as high as 4.65 g/kg DM does not negatively affect growth performance if AAs are balanced.

In a feeding trial with weaned piglets, 10 and 20% of faba beans with low (< 0.05 g/kg DM) and high tannin content (4.65 g/kg DM) were included. In both groups, average daily feed intake, weight gain and feed conversion ratio did not differ from the control diet without faba beans. It was concluded that, regardless of tannin content, faba beans can be included in the diet of weaned piglets at levels up to 20%. (table 2.5.3.2-1; Royer *et al.*, 2010). In another feeding trial even 35% of tannin rich

TABLE 2.5.3.2-1 Recommended feeding levels for faba beans (*Vicia faba L.*) of unknown origin for pigs (% of diet)

Animal group	Body weight / reproductive cycle	Faba beans % of diet	Essential AA[1] balanced diet	Authors
Piglets	8–30 kg	10	–[2]	Crépon *et al.*, 2010
		20	+[3]	Royer *et al.*, 2010
Fattening pigs	30–60 kg	30	+	Smith *et al.*, 2013
Sows	pregnant	15	*[4]	Hoffmann, Steinhöfel, 2023
	lactating	16	*	Świątkiewicz *et al.*, 2018

1 AA: amino acid.
2 – No.
3 + Yes.
4 * Not stated.

faba beans had no adverse effects on average daily feed intake, weight gain and feed conversion ratio in a diet well balanced in essential AAs (Royer *et al.*, 2010). However, without the addition of methionine and tryptophan, a diet should not contain more than 10% of faba beans. Otherwise daily weight gain and feed conversion ratio may be reduced (Crépon *et al.*, 2010).

In a pig fattening trial, the colour-flowered and high tannin variety 'Fuego' was integrated into to diet from 7.5% up to 30%, gradually and finally completely replacing soybean meal (Smith *et al.*, 2013). The diet was adjusted for net energy and digestible lysine levels and met the minimum requirements of Met, Thr, Trp, Ca, and digestible P. In the growing period (30 to 60 kg body weight), there was no significant difference in feed intake, daily gain and feed conversion ratio when compared to soybean meal. However, finisher pigs (60 to 100 kg) had lower daily weight gain in all groups, independent of the diet. Since grower and finisher were not the same pigs, there may have been an individual difference in daily weight gain. Feed intake and feed conversion ratio were not affected. Carcass quality (back fat thickness at P2, percent of lean meat) were also not affected by the diet, but the percentage of killing out was lower in the group with the highest amount of faba beans. Boar taint inducing skatole concentrations were also not different between the soybean meal group and any of the faba bean groups (Smith *et al.*, 2013). Moreover, they were well below the accepted threshold level of 0.2 µg/g back fat for boar taint detection. Indole concentrations tended to decrease with higher faba bean levels in the diet. This may be due to the high raffinose and stachyose content in legumes which are fermented in the hind gut. There were no effects on carcass quality or back fat skatole concentrations, but indole concentration was linearly reduced with increasing legume inclusion ($p = 0.05$). The results of this study suggest that a much higher level of faba beans can be incorporated into pig diets than is usually assumed, as long as they are balanced for essential amino acids.

A feeding trial with sows, receiving 12% of faba beans with low or high tannin content (0.68 and 8.86 g/kg, respectively) during pregnancy and 16% during lactation revealed no differences in reproductive traits, litter weight or body weight of new born piglets compared to groups fed with soybean meal (Świątkiewicz *et al.*, 2018). Hoffmann and Steinhöfel (2023) recommend levels up to 15% for either stage of the reproductive cycle (table 2.5.3.2-1).

2.5.3.3 Poultry

2.5.3.3.1 *Broiler*

In faba beans, the content of metabolisable energy is lower than in cereals but higher than in soybean meal. Tannins and vicine and convicine (V/C) are the main antinutritive factors with regard to poultry feeding. In an extensive feeding trial with broilers, soybean meal and wheat were replaced by four varieties of faba beans at concentration of 12% (d 0–12), 24% (d 13–25) and 35% (d 26–41). Faba beans were either low or high in tannin (< 0.05 and 5.2 to 6.1 g/kg) or V/C (0.7 and 12.4 to 13.2 g/kg) content. Essential amino acids were added to meet the requirements. No negative effect of tannins and V/C in feed intake was observed. Average daily weight gain was not significantly different between the diets in any feeding period, but was lower in the V/C containing varieties when the total feeding period was considered. Over the entire feeding period, weight gain was also lower in all experimental diets compared to the control diet containing soybean meal. Weight gain and feed conversion ratio were lower in the finisher feeding period with 36% of faba beans and throughout the feeding period for all experimental diets compared to the control, but not within feeding periods. However, there was also a significant cultivar effect on total daily gain, with tannin and V/C content appearing to have an additive effect. On the other hand, there was also a cultivar effect on feed intake that could not be explained by tannin or V/C content. Dehulling faba beans and therefore reducing the tannin and fibre content had no effect on broiler growth performance, chilled carcass weight, dressing percentage, and yield of other saleable cuts. Therefore, faba beans can be readily fed to broiler chickens without the need for dehulling. The authors consider the V/C content of faba beans more important than the tannin content in broiler diet formulation (Cho *et al.*, 2019).

In laying hens, the growing conditions of faba beans and their phytate content are more important for amino acid digestibility than the V/C content. Digestibility is also influenced by the tannin content. Tannins reduce overall digestibility of organic matter and thus the energy content (Rodehutscord and Sieger, 2020). A feeding trial with V/C rich and poor faba bean varieties was carried out. At a rate of

TABLE 2.5.3.3-1 Recommendations for the inclusion of faba beans (*Vicia faba L.*) of unknown origin in poultry feed (% of diet)

Poultry group	**Faba beans (White-flowered)**	**Faba beans (Colour-flowered)**
Laying hens		
Egg production	10	10
Reproduction	5	5
Broiler		
Starter (up to week 4)	20	10
Grower/finisher	25	20
Turkey		
Starter (P1/P2)[1]	15	5/10
Grower (P3–4)	20	15
Finisher (P5–7)	20	15

1 Production period.
BELLOF *ET AL.*, 2020

15 and 20% in the diet, V/C rich faba beans reduced feed intake, eggs per hen/per day and egg mass produced per day. V/C poor varieties at inclusion levels up to 37% had no negative effect on daily feed intake, feed conversion ratio, eggs per hen/per day, egg mass and egg quality (Halle, 2020).

The effect of substituting soybean meal with faba beans was also tested in laying hens. At the inclusion of 10% faba beans, Halle (2005) registered a slight reduction in feed intake and in daily produced egg mass even with a cultivar low in tannin but high in V/C (9.1 g/kg) content. At the rate of 20% in the diet, both effects were significant. Therefore, restrictions in the substitution of soybean meal with faba beans of unknown origin and cultivar is still recommended (table 2.5.3.3-1).

2.5.3.4 Horses

Horses with higher energy and/or protein requirements (growing, lactating, heavy work) are fed with concentrates. Main ingredients of these concentrates are oats and soybean meal, main nutrients starch and protein. Both are also the main nutrients in faba beans and it was tested to what extent oats and soybean meal could be substituted. The substitution of soybean meal is desired for ethical reasons, as most of it is imported into the EU and rainforest is cut down in Brazil to grow soya. Oats, on the other hand, are said to cause agitation in some horses. It may therefore be desirable to replace as much of these ingredients as possible.

Faba bean cultivars with small seeds are also known as horse beans. This suggests that they were historically used in equine nutrition. In the past they were valued as a source of energy for horses because of their starch content. However, the tannin in faba beans can cause constipation. In addition, high concentrations of fibre that are digested in the large intestine can cause bloating. Therefore, only small amounts of faba beans should be fed to horses. Coenen (2020) recommends limiting faba beans to 0.1 or 0.2 kg/100 kg body weight per day. In crib feed, a concentration of 15% can be included (Hoffmann and Steinhöfel, 2023).

2.5.3.5 Fishes and shrimps

Vanessa Fuchs, Monika Weiß and Sinem Zeytin

Faba beans (*Vicia faba L.*) are a suitable aquafeed ingredient, which is high in protein (around 20–40%), energy and carbohydrates (predominantly starch). In aquafeeds, faba bean (FB) products are used as protein sources, partially substituting fishmeal or soybean products, or as an additive. Additionally, the carbohydrates in faba bean meal (FBM) have favourable binding properties, ensuring good product quality of extruded feed pellets. Thus, they can also be utilised to replace wheat as a starch source in diets. The disadvantage of raw FB is that it contains a variety of antinutritive factors (ANF), including different enzyme inhibitors, phytic acid, tannins and lectins. Vicine and convicine are the two major antinutritive compounds found in faba beans. Antinutrients may negatively affect feed palatability and intake, the bioavailability of protein and energy, nutrient digestibility and animal performances. For more information on ANF refer to chapter 2.3.

The dehulling of beans may be beneficial before using them for aquafeeds, especially at high dietary inclusion levels. This process increases the protein content, reduces the fibre content and removes a large proportion of the ANFs. It is also recommended to inactivate ANFs by heat processing (roasting, autoclaving, microwaving, cooking or extruding) or by other methods such as soaking, germination or fermentation, prior to inclusion in fish feeds. Moreover, air classification is a process, which is applied to obtain a product with a high protein concentration (chapter 2.9.1). In feeds for piscivorous species, a protein concentrate with protein levels > 55% is preferably used to fulfil the high dietary protein requirement. However, air classification during industrial scale processing has been shown to increase ANFs of the product as no heating is applied.

To date, FB products have been tested in feed for some fish species, but not for shrimp species. In studies investigating the digestibility of raw FB, high nitrogen/protein digestibility (> 90%) was noted for Silver perch (*Bidyanus bidyanus*) and Nile tilapia (*Oreochromis niloticus*) (Allan *et al.*, 2000; Booth *et al.*, 2001; Magalhães *et al.*, 2018). 30% dehulled FB was shown to increase the protein or amino acid digestibility of silver perch diets compared to whole FB (Booth *et al.*, 2001). Protein digestibility showed to be lower (> 80%) for Rainbow trout (*Oncorhynchus mykiss*) and Atlantic salmon (*Salmo salar*) (Gomes *et al.*, 1995; Aslaksen *et al.*, 2007; Magalhães *et al.*, 2018). In general, starch digestibility of legumes, like FB, is lower than that of cereals because of the higher gelatinisation temperature and higher resistance to hydrolysis of bean starch.

In diets for Beluga (*Huso huso*) juveniles, 10% FBM from whole beans, partly replacing fishmeal, can be included without adverse effects on growth performance, digestibility, flesh quality and health (Soltanzadeh *et al.*, 2016, Soltanzadeh *et al.*, 2017). Diets for juvenile Nile tilapia can contain 24% FBM by replacing dehulled soybean meal (SBM) without any negative effects on fish performance and protein digestibility (Azaza *et al.*, 2009). Higher inclusion of 37% FBM may be applied using cooked decorticated waste seeds in diets for Nile tilapia without affecting growth and nutrient utilization (Gaber, 2006).

Grass carp (*Ctenopharyngodon idella*) can accept 42% dietary FBM showing high growth gain and feed utilization compared to a diet with high SBM content (Gan *et al.*, 2017; Li *et al.*, 2020). However, such a high level decreased digestive enzyme activities, lipid content and health in fish.

On the other hand, feeding FB improved the textural flesh quality (hardness, chewiness and springiness) in Nile tilapia, Grass carp and Common carp (*Cyprinus carpio*). It is assumed that the antinutrient vicine, which is a vital substance in FB, could be responsible for the improved textural quality. Some studies revealed a positive effect on muscle fibres, fillet texture in terms of "crispiness" and collagen content as well as increased amino and fatty acids in Nile tilapia at a dietary level of 60% FBM (Peng *et al.*, 2022; Li *et al.*, 2023a; Li *et al.*, 2023b). Similarly, dietary protein replacement with 63% FBM inclusion improved flesh quality of Grass carp and is suggested in the cultivation of crisp grass carp (Fu *et al.*, 2022). FBM can also improve the fillet quality for Common/Yellow River carp, but in high amounts may lead to poor growth performance and fish health (Song *et al.*, 2020).

In addition, FB extracts were tested as a feed additive in Grass carp diets. A FB water extract was considered to have high potential in improving the textural quality with less negative effects on fish performances and health. In diets for Grass carp, 13.6% FB water extract promoted growth, improved the intestinal microbiota and health (Ma *et al.*, 2022). Similarly, using 1.25% FB water extract in *Nibea coibor* diets may have beneficial effects on growth, health and textural quality of the swim bladder by promoting collagen deposition (Guo *et al.*, 2024).

Extruded FB can be included in European seabass (*Dicentrarchus labrax*) and Gilthead seabream (*Sparus aurata*) diets up to 35% substituting for wheat, SBM and sunflower meal with no negative effects on growth performance (Adamidou *et al.*, 2009; Adamidou *et al.*, 2011). However, extruded FB reduced the total fat content of gilthead seabream.

Rainbow trout fingerlings can tolerate up to 30% dehulled FBM in diets, but maximum growth performance and higher protein digestibility was found at the 15% FBM inclusion level (Ouraji *et al.*, 2013). Up to 12% air classified FB concentrate (62% protein) in diets for Atlantic salmon parr, partially replacing soy protein concentrate and fishmeal, resulted in high weight gain and reduced incidence of intestinal inflammation compared to diets with high content of soy protein concentrate

TABLE 2.5.3.5-1 Recommended inclusion levels of faba beans in feed mixtures of different fish species.

Species	Initial species weight (g)	Product	Replaced resource	Inclusion rates (%; unless otherwise specified)	Recommendable inclusion rate (%)	Reference s. App. 1
Beluga (*Huso huso*)	~82	Faba bean meal	Fishmeal	5, 10, 15, 20 and 25	10	Soltanzadeh *et al.*, 2016
	~82	Whole faba bean seed meal, untreated	Fishmeal	5, 10, 15, 20 and 26	10	Soltanzadeh *et al.*, 2017
Nile tilapia (*Oreochromis niloticus*)	~17.3	Faba bean meal, untreated	Dehulled soybean meal	12, 24 and 36	Up to 24	Azaza *et al.*, 2009
	~1.9	Broad (faba) bean meal, decorticated waste seed and boiled for 30 min.	Fishmeal	18.4, 37, 55 and 80	37	Gaber, 2006
	~463.9	Faba bean meal	Soybean meal, rapeseed meal and wheat flour	40, 50, 60 and 70	60	Li, Huang, Zhang, Qin, *et al.*, 2023a
	~463.9	Faba bean meal	Soybean meal, rapeseed meal and wheat flour	40, 50, 60 and 70	–	Li, Huang, Zhang, Zou, and Lin, 2023b
	~500.2	Faba bean meal	Soybean meal and rapeseed meal	0, 15, 30 and 60	Up to 60	Peng *et al.*, 2022

TABLE 2.5.3.5-1 Recommended inclusion levels of faba beans in feed mixtures of different fish species (*cont.*)

Species	Initial species weight (g)	Product	Replaced resource	Inclusion rates (%; unless otherwise specified)	Recommendable inclusion rate (%)	Reference s. App. 1
Grass carp (*Ctenopharyngodon idella*)	~3.4	Faba bean meal (whole seed)	Soybean meal	14, 28, 42 and 56	42	Gan *et al.*, 2017
	~3.4	Faba bean meal (whole seed)	Soybean meal	14, 28, 42 and 56	42	Li *et al.*, 2020
	~130.1	Faba bean meal	Soybean meal, rapeseed meal and wheat flour	63, 72, 81 and 90	63	Fu *et al.*, 2022
	~750	Whole faba bean (FB) and four FB extracts (FB water extract, FB alcohol extract, FB proteins, and FB residues)	–	FB[a] alcohol extract: 0.108 kg; FB water extract: 2.031 kg; FB proteins: 4.783 kg; FB residues: 7.989 kg	FB water extract	Ma *et al.*, 2020
	~838	Whole faba bean (FB) and four FB extracts (FB water extract, FB alcohol extract, FB proteins, and FB residues)	–	FB water extract: 13.62; FB alcohol extract: 0.72; FB proteins: 32.08; FB residues: 33.58	FB water extract	Ma *et al.*, 2022
Common carp (*Cyprinus carpio*)	~500	Broken faba bean	–	Faba bean pieces: 100	–	Song *et al.*, 2020

a Faba bean.

TABLE 2.5.3.5-1 Recommended inclusion levels of faba beans in feed mixtures of different fish species (*cont.*)

Species	Initial species weight (g)	Product	Replaced resource	Inclusion rates (%; unless otherwise specified)	Recommendable inclusion rate (%)	Reference s. App. 1
Nibea coibor	~8.9	Additive: faba bean (FB) water extract or FB alcohol extract	–	FB water extract: 1.25, 2.5, and 5; FB alcohol extract: 0.9, 1.8, and 3.6	FB water extract: 1.25	Guo *et al.*, 2024
European seabass (*Dicentrarchus labrax*)	Feeding trial: ~97.9, digestibility trial: ~176.5	Meal, extruded	Soybean, sunflower and wheat meal	16.5 and 34	Up to 34	Adamidou *et al.*, 2009
Gilthead seabream (*Sparus aurata*)	~92.6	Meal, extruded	Soybean, sunflower and wheat meal	17.5 and 35	Up to 35	Adamidou *et al.*, 2011
Rainbow trout (*Oncorhynchus mykiss*)	~12.5	Dehulled faba bean meal	Soybean meal, fishmeal	15, 30 and 45	15 and 30 (but less growth)	Ouraji *et al.*, 2013

TABLE 2.5.3.5-1 Recommended inclusion levels of faba beans in feed mixtures of different fish species (*cont.*)

Species	Initial species weight (g)	Product	Replaced resource	Inclusion rates (%; unless otherwise specified)	Recommendable inclusion rate (%)	Reference s. App. 1
Atlantic salmon (*Salmo salar*) parr	~1.5	Concentrate, dried by air-classification	Soy protein concentrate	11.2, 22.4, 33.5 and 44.7	12	De Santis, Crampton, Bicskei, and Tocher, 2015a
	~1.5	Concentrate, dried by air-classification	Soy protein concentrate, fishmeal	6.5 to 58.5	11.7	De Santis, Ruohonen, *et al.*, 2015b
Atlantic salmon post-smolt	~204	Wet processed faba bean isolate	Soy protein concentrate, fishmeal	7 and 14	7	De Santis *et al.*, 2016a
	~1276	Concentrate, dried by air-classification	Soy protein concentrate, fishmeal	10 and 21	21	De Santis *et al.*, 2016b

(De Santis *et al.*, 2015a; De Santis *et al.*, 2015b). In post-smolt salmons, a higher inclusion level of 21% air classified FB concentrate was successfully applied in combination with soy protein concentrate and fishmeal (De Santis *et al.*, 2016b). Mixed plant protein diets containing FBM and soy protein concentrate appear to have a positive effect on salmon performance and health compared to single plant protein diets with either FB concentrate or soy protein concentrate alone. This is probably due to reduced levels of individual ANFs.

A wet processed protein isolate is another suitable FB product, which can be included at a level of 7% in diets for post-smolt Atlantic salmon reducing FM and soy protein concentrate (De Santis *et al.*, 2016a). However, this low amount already caused mild intestinal inflammation in salmons.

References

Adamidou S., Nengas I., Henry M., Grigorakis K., Rigos G., Nikolopoulou D., Kotzamanis Y., Bell G.J., Jauncey K., 2009: Growth, feed utilization, health and organoleptic characteristics of European seabass (*Dicentrarchus labrax*) fed extruded diets including low and high levels of three different legumes. Aquaculture, 293, 263–271.

Adamidou S., Nengas I., Henry M., Ioakei Midoy N., Rigos G., Bell G., Jauncey K., 2011: Effects of dietary inclusion of peas, chickpeas and faba beans on growth, feed utilization and health of gilthead seabream (*Sparus aurata*). Aquaculture nutrition, 17, e288–e296.

Al Shabuol N.L., Obeidat B.S., 2022: Feeding of faba beans (*Vicia faba* L.) enhances the growth performance of lambs. Vet. World, 15, 906–910.

Allan G.L., Parkinson S., Booth M.A., Stone D.A.J., Rowland S.J., Frances J., Warner-Smith R., 2000: Replacement of fish meal in diets for Australian silver perch, *Bidyanus bidyanus*: I. Digestibility of alternative ingredients. Aquaculture, 186, 293–310.

Aslaksen M.A., Kraugerud O.F., Penn M., Svihus B., Denstadli V., Jørgensen H.Y., Hillestad M., Krogdahl Å., Storebakken T., 2007: Screening of nutrient digestibilities and intestinal pathologies in Atlantic salmon, *Salmo salar*, fed diets with legumes, oilseeds, or cereals. Aquaculture, 272, 541–555.

Azaza M.S., Wassim K., Mensi F., Abdelmouleh A., Brini B., Kraïem M.M., 2009: Evaluation of faba beans (*Vicia faba* L. var. minuta) as a replacement for soybean meal in practical diets of juvenile Nile tilapia *Oreochromis niloticus*. Aquaculture, 287, 174–179.

Bellof G., Freitag M., 2021: Körnerleguminosen. Editors: Durst L., Freitag M., Bellof G, 2021: Futtermittel für landwirtschaftliche Nutztiere. DLG Verlag Frankfurt/Main, 258–304.

Bellof G., Halle I., Rodehutscord M., 2020: UFOP-Praxisinformationen: Ackerbohnen, Körnerfuttererbsen, Süßlupinen und Sojabohnen in der Geflügelfütterung. Editor: Union zur Förderung der Öl- und Proteinpflanzen e.V. (UFOP). Berlin. 2nd updated edition.

Bello-Pérez L.A., Islas-Hernández J.J., Rendón-Villalobos J.R., Agama-Acevedo E., Morales-Franco L., Tovar J., 2007: In vitro starch digestibility of fresh and sun-dried faba beans (*Vicia faba L.*). J. Sci. Food and Agriculture, 87, 1517–1522.

Bissinger C., Schneider K., Steingass H., 2007: Untersuchungen an Milchkühen zur Erhöhung der nXP-Versorgung bei Einsatz von Ackerbohnen, Erbsen und Lupinen. Heimische Körnerleguminosen mit geschütztem Protein in der Milchviehfütterung. UFOP publication series, 33, 21–27.

Booth M.A., Allan G.L., Frances J., Parkinson S., 2001: Replacement of fish meal in diets for Australian silver perch, *Bidyanus bidyanus*: IV. Effects of dehulling and protein concentration on digestibility of grain legumes. Aquaculture, 196, 67–85.

Caracuta V., Barzilai O., Khalaily H., Milevski I., Paz Y., Vardi J., Regev L., Boaretto E., 2015: The onset of faba bean farming in the Southern Levant. Scientific reports, 5, 14370. doi: 10.1038/srep14370. ISSN 2045-2322.

Cho M., Smit M.N., He L., Kopmels F.C., Beltranena E., 2019: Effect of feeding zero- or high-tannin faba bean cultivars and dehulling on growth performance, carcass traits and yield of saleable cuts of broiler chickens. J. Appl. Poult. Res., 28,1305–1323. http://dx.doi.org/10.3382/japr/pfz099.

Coenen M., 2020: Futtermittel. Editors: Coenen M. and Vervuert I., 2020: Pferdefütterung. Georg Thieme Publisher. Stuttgart – New York. 6th revised edition.

Crépon K., Marget P., Peyronnet C., Carrouée B., Arese P., Duc G., 2010: Nutritional value of faba bean (*Vicia faba L.*) seeds for feed and food. Review. Field crops research, 115, 329–339.

CVB, 2018: Veevoedertabel – Chemische samenstellingen en nutritionele waarden van voedermiddelen. Productschap Diervoeder, Den Haag.

De Santis C., Crampton V.O., Bicskei B., Tocher D.R., 2015a: Replacement of dietary soy – with air classified faba bean protein concentrate alters the hepatic transcriptome in Atlantic salmon (*Salmo salar*) parr. Comp biochem physiol part D genomics Proteomics, 16, 48–58.

De Santis C., Ruohonen K., Tocher D.R., Martin S.A.M., Król E., Secombes C.J., Bell J.G., El-Mowafi A., Crampton V.O., 2015b: Atlantic salmon (*Salmo salar*) parr as a model to predict the optimum inclusion of air classified faba bean protein concentrate in feeds for seawater salmon. Aquaculture, 444, 70–78.

De Santis C., Martin S.A.M., Dehler C.E., Iannetta P.P.M., Leeming D., Tocher D.R., 2016a: Influence of dietary inclusion of a wet processed faba bean protein isolate on post-smolt Atlantic salmon (*Salmo salar*). Aquaculture, 465, 124–133.

De Santis C., Tocher D.R., Ruohonen K., El-Mowafi A., Martin S.A.M., Dehler C.E., Secombes C.J., Crampton V., 2016b : Air-classified faba bean protein concentrate is efficiently utilized as a dietary protein source by post-smolt Atlantic salmon (*Salmo salar*). Aquaculture, 452, 169–177.

Dhull S.B., Kidwai M.K., Noor R., Chawla P., Rose P.K., 2021: A review of nutritional profile and processing of faba bean (*Vicia faba L.*). Legume science, e129. wileyonlinelibrary.com/journal/legumescience.

Föste M., 2021: Erhöhung der ernährungsphysiologischen Qualität von Ackerbohnenmehlen und -konzentraten durch Reduktion antinutritiver Inhaltsstoffe. Abschlussbericht zum BLE geförderten Projekt, Förderkennzeichen 2815EPS011: 76.

Fu S., Wang B., Zhu Y., Xue Y., Zhong W., Miao Y., Du Y., Wang A., Wang L., 2022: Effects of faba bean (*Vicia faba*) diet on amino acid and fatty acid composition, flesh quality and expression of muscle quality-related genes in muscle of "crispy" grass carp, *Ctenopharyngodon idella*. Aquaculture research, 53, 4653–4662.

Gaber M.M., 2006: Partial and complete replacement of fish meal by broad bean meal in feeds for Nile tilapia, *Oreochromis niloticus* L., fry. Aquaculture research, 37, 986–993.

Gan L., Li X.X., Pan Q., Wu S.L., Feng T., Ye H., 2017: Effects of replacing soybean meal with faba bean meal on growth, feed utilization and antioxidant status of juvenile grass carp, *Ctenopharyngodon idella*. Aquaculture nutrition, 23, 192–200.

Gdala J., Buraczewska L., 1997: Chemical composition and carbohydrate content of several varieties of faba bean and pea seeds. J. Anim. Feed sci., 6, 123–135.

Gomes E.F., Rema P., Kaushik S.J., 1995: Replacement of fish meal by plant proteins in the diet of rainbow trout (*Oncorhynchus mykiss*): digestibility and growth performance. Aquaculture, 130, 177–186.

Guo H., Liu G., Tian M., Liu C., Zhang H., Wang S., Wen X., Lin F., 2024: Effect of faba bean *Vicia faba* L. water/alcohol extract on growth performance, antioxidant capacity, textural properties, and collagen deposition in the swim bladder of juvenile *Nibea coibor*. Fish physiology and biochemistry, 50, 1157–1169.

Halle I., 2005: Einfluss gestaffelter Anteile von je zwei Erbsen- und Ackerbohnensorten im Legehennenfutter auf die Leistungsmerkmale. Landbauforschung Völkenrode, 3, 149–155.

Halle I., 2020: Züchtung und Agronomie neuartiger, Vicin-armer Ackerbohnen und Einsatz als einheimisches Eiweißfutter. BLE gefördertes Forschungsprojekt, Abschlussbericht. Förderkennzeichen: FKZ: 2815EPS063.

Hans M., Kaur G., Salaria A., Sravan T., 2021: Agrarian conditions and post-harvest practices of faba bean. In: Faba bean: chemistry, properties and functionality (Editors: Bangar S.P., Dhull S.B.). Springer, ISBN 978-3-031-14587-2 (eBook). doi: 10.1007/978-3-031-14587-2.

Hoffmann M., Steinhöfel O., 2023: Futtermittel- und rationsspezifische Restriktionen. Agrimedia, Erling Verlag GmbH & Co. KG.

Johnston D.J., Theodoridou K., Gordon A.W., Yan T., Mcroberts, W.C., Ferris C.P., 2019: Field bean inclusion in the diet of early-lactation dairy cows: effects on performance and nutrient utilization. J. Dairy sci., 102, 10887–10902.

LfL Bayern, 2017: Gruber Tabelle zur Fütterung der Milchkühe, Zuchtrinder, Schafe, Ziegen. Bayerische Landesanstalt für Landwirtschaft, 42nd edition.

Li Q., Huang Y., Zhang X., Qin Z., Zou C., Tan X., Xie X., Liang S., Lin L., 2023a: Effects of faba beans (*Vicia faba* L.) on growth performance, textural quality and physiological indices of tilapia (*Oreochromis niloticus*). Aquaculture, 574, 739640.

Li Q., Huang Y., Zhang X., Zou C., Lin L., 2023b: Improvement of muscle quality in tilapia (*Oreochromis niloticus*) with dietary faba bean (*Vicia faba* L.). Frontiers in nutrition, 10, 1153323.

Li X., Chen S., Sun J., Huang X., Tang H., He Y., Pan Q., Gan L., 2020: Partial substitution of soybean meal with faba bean meal in grass carp (*Ctenopharyngodon idella*) diets, and the effects on muscle fatty acid composition, flesh quality, and expression of myogenic regulatory factors. Journal of the world aquaculture society, 51, 1145–1160.

Ma L.-L., Zhang J.-M., Kaneko G., Xie J., Sun J.-H., Wang G.-J., Tian J.-J., Zhang K., Li Z.-F., Gong W.-B., Xia Y., Yu E.-M., 2022: Growth performance, intestinal microbiota and immune response of grass carp fed isonitrogenous and isoenergetic diets containing faba bean extracts. Aquaculture reports, 22, 100924.

Magalhães S.C.Q., Cabrita A.R.J., Valentão P., Andrade P.B., Rema P., Maia M.R.G., Valente L.M.P., Fonseca A.J.M., 2018: Apparent digestibility coefficients of European grain legumes in rainbow trout (*Oncorhynchus mykiss*) and Nile tilapia (*Oreochromis niloticus*). Aquaculture nutrition, 24, 332–340.

Meng Z., Liu Q., Zhang Y., Chen J., Sun Z., Ren C., Zhang Z., Cheng X., Huang Y., 2021: Nutritive value of faba bean (*Vicia faba L.*) as a feedstuff resource in livestock nutrition: a review. Food sci. Nutr., 9, 5244–5262.

Offner A., Bach A., Sauvant D., 2003: Quantitative review of in situ starch degradation in the rumen. Anim. Feed sci. Technol., 106, 81–93.

Ouraji H., Zaretabar A., Rahmani H., 2013: Performance of rainbow trout (*Oncorhynchus mykiss*) fingerlings fed diets containing different levels of faba bean (*Vicia faba*) meal. Aquaculture, 416–417, 161–165.

Peng K., Fu B., Li J., Zhao H., Cao J., Huang W., Chen B., Li X., Peng Z., Wei M., 2022: Effects of replacing soybean meal and rapeseed meal with faba bean meal on growth performance and muscle quality of tilapia (*Oreochromis niloticus*). Aquaculture Reports, 26, 101328.

Rodehutscord M., Siegert W., 2020: Züchtung und Agronomie neuartiger, Vicin-armer Ackerbohnen und Einsatz als einheimisches Eiweißfutter. Aminosäurenverdaulichkeit und Umsetzbare Energie bei Legehennen. Abschlussbericht des BLE geförderten Projekts, Förderkennzeichen: 2815EPS065.

Royer E., Crépon K., Granier R., Peyronnet C., Vilariño M., 2010: Incidence du type de féverole et du taux d'incorporation sur les performances du porc en post-sevrage et engraissement. Journées Recherche Porcine, 77–83.

Smith L.A., Houdijk J.G.M., Homer D., Kyriazakis I., 2013: Effects of dietary inclusion of pea and faba bean as a replacement for soybean meal on grower and finisher pig performance and carcass quality. J. Anim. Sci., 91, 3733–3741.

Soltanzadeh S., Esmaeili Fereidouni A., Ouraji H., Khalili K.J., 2016: Growth performance, body composition, hematological, and serum biochemical responses of beluga (*Huso huso*) juveniles to different dietary inclusion levels of faba bean (*Vicia faba*) meal. Aquaculture International, 24, 395–413.

Soltanzadeh S., Ouraji H., Fereidouni A.E., Ahmadi H.A.S., Khalesi M.K., 2017: Growth, feed utilization, apparent digestibility and fatty acid composition of beluga (*Huso huso*) juveniles to different inclusion levels of whole seed meal of faba bean (*Vicia faba*). Turkish journal of fisheries and aquatic sciences, 17, 1415–1424.

Song D., Yun Y., Mi J., Luo J., Jin M., Nie G., Zhou Q., 2020: Effects of faba bean on growth performance and fillet texture of Yellow River carp, *Cyprinus carpio* haematopterus. Aquaculture reports, 17, 100379.

Świątkiewicz M., Księżak J., Hanczakowska E., 2018: The effect of feeding native faba bean seeds (*Vicia faba L.*) to sows and supplemented with enzymes to piglets and growing pigs. Ann. Anim. Sci., 18, 1007–1027.

2.6 Sweet lupins (*Lupinus sp.*)

Mechthild Freitag

Lupins appear to have been domesticated later than peas or beans. The centers of domestication were in Asia Minor and the Balkans, from where they spread to the Mediterranean and North Africa. Only three species are cultivated in Europe: the white- (*Lupinus albus*), yellow- (*Lupinus luteus*) and narrow-leaved (*Lupinus angustifolius*) lupin. A second center of domestication with different species was established in America.

In northern Europe in the 19th century, lupins were mainly grown as green manure because of their ability to bind atmospheric nitrogen. Their use in animal feed was limited because of their high alkaloid concentrations and low seed yield. In Germany, widespread cultivation only began in the 20th century, when

FIGURE 2.6-1
L. angustifolius, sprouting
PHOTO: DSV

FIGURE 2.6-2
L. angustifolius, flowering
PHOTO: OYOYOY, CC BY-SA 3.0 WIKIPEDIA

FIGURE 2.6-3
White-flowered lupin "Frieda"

FIGURE 2.6-4
Lupin plant with pods

yellow- and narrow-leafed lupins with low alkaloid concentrations were discovered and bred. Even today, however, lupins are used less in animal feed than peas or beans (Pastuszewska, 2016).

2.6.1 *Ingredients*

2.6.1.1 Nutrients and antinutritive factors

2.6.1.1.1 *Nutrients*

Lupins can be divided into two categories: those with a high alkaloid content (up to 53.8 g/kg), commonly known as bitter lupins, and those with a low alkaloid level (less than 0.5 g/kg), known as sweet lupins (David *et al.*, 2024). Only the latter can be used for feeding purposes. The quinolizidine alkaloids like lupinine, sparteine, lupanine and others in bitter lupins cause animals to reject the seeds. For feeding purposes, the alkaloid level has been reduced to less than 0.1 g/kg and to less than 0.05 g/kg if larger quantities are included in the diet. This has been achieved by breeding. However, the low alkaloid level makes the plants more susceptible to disease and parasites. In addition, the alkaloid content can increase under unfavourable growing conditions such as drought. For more information on the effects of alkaloids, see section 2.3.1.

Due to the fungus anthracnose, yellow-flowered species have not been cultivated in Germany since 1995. Because of their high protein level, they would be interesting for feeding purposes and it is to be hoped that currant breeding programmes will be successful in cultivating this species again. Sweet narrow-leafed species and newly bred varieties of white-flowered sweet lupins are more resistant to anthracnose.

In comparison to peas and faba beans, lupins and especially white-flowered varieties have a higher protein content and for animal nutrition purposes a better balanced amino acid profile. They have no starch and low levels of sugar. But they contain more fat than peas and faba beans with a high level of unsaturated fatty acids, especially linoleic acid (table 2.6.1.1-1).

TABLE 2.6.1.1-1 Ingredients (g/kg DM) in narrow-leafed (*L. angustifolius*), white- (*L. albus*) and yellow- (*L. luteus*) flowered sweet lupins

Ingredient	***L. angustifolius***			***L. albus***		***L. luteus***	
			Min.–max.				
DM	880	1000	1000	880	1000	880	1000
CP	289	328	180–330	339	385	419	476
CL	56	64	42–65	83	95	50	57
CF	141	160	110–171	113	129	139	158
$aNDF_{om}$	220	250	150–240	167	190	222	252
ADF_{om}	180	205	140–240	128	145	183	208

TABLE 2.6.1.1-1 Ingredients (g/kg DM) in narrow-leafed (*L. angustifolius*) (*cont.*)

Ingredient	***L. angustifolius***			***L. albus***		***L. luteus***	
			Min.–max.				
STA[1]	70	80	11–170	77	88	70	80
SUG	50	28	20–70	64	72	9	10
NFE	359	408	*	295	260	295	260
ASH	35	40	30–50	35	40	43	49

1 Method related evidence.
* No data.
BELLOF AND FREITAG, 2021

FIGURE 2.6.1-1
Lupin pod

FIGURE 2.6.1-2
Lupin pod, open
PHOTO: DSV

2.6.1.1.2 *Carbohydrates*

In general, the fibre content of lupins is higher than that of peas or faba beans. Together with the level of fat, they ensure a high energy value for ruminants. For non-ruminants, especially for young poultry, the amount of non-starch polysaccharids (NSP) of approximately 389 g/kg DM in narrow-leafed and 358 g/kg in white-flowered cultivars may be a disadvantage in the diet.

2.6.1.1.3 *Antinutritive factors (ANF)*

For a more detailed description of ANF in legumes refer to section 2.3.

In general, lupins have lower levels of undesirable components such as phytic acid, oligosaccharides, trypsin inhibitors, lectins and saponins compared to other legumes (Kohajdova *et al.*, 2011). Unlike faba beans and peas, lupins are virtually free of phenolic compounds. They contain less than one percent of the lectin

TABLE 2.6.1.1-2 Carbohydrate level (g/kg DM) in *L. angustifolius* and *L. albus*

Parameter	*L. angustifolius*		*L. albus*	
	Mean	Range	Mean	Range
Total fibre	*	*	*	344–394
CF	179	140–213	134	105–162
NDF	272	240–307	203	185–234
ADF	224	198–258	158	130–172
Soluble fibre	34	*	44	36–52
Insoluble fibre	488	*	325	308–342
STA	80	11–170	50	14–125
SUG	57	20–70	72	*

* No data.
BELLOF AND FREITAG, 2021; DAVID *ET AL.*, 2024

activity of faba beans and peas. Saponins may be present and the highest levels have been found in narrow-leafed lupins, ranging from 0.379 to 0.740 g/kg (Jezierny *et al.*, 2010). These levels are very low, compared to soya and faba beans.

The main ANFs that reduce lupin inclusion in animal diets are alkaloids, oligosaccharides and non-starch polysaccharides (NSP). Of these, various alkaloids such as lupinine, lupanine, sparteine, angustifoline and, to a lesser extent, lupinidine, hydroxilupanine, anagirine, monolupin, termophsine, and puziline are the most important (Kohajdova *et al.*, 2011). The predominant alkaloid in *L. albus* seeds is lupanine, followed by 13-OH-lupanine, angustifoline and isolupanine, whereas lupinine is the dominant alkaloid in *L. luteus*. Alkaloids can be toxic and cause liver damage. The alkaloid content of bitter narrow-leafed lupins can be 1000 times greater than that of sweet lupins. Bitter lupin varieties have a higher seed yield and are better at repelling insects, thus making them more attractive for cultivation but unsuitable for feeding. The alkaloid content of bitter *L. albus* cultivars ranges from 5 to 40 g/kg, while that of low-alkaloid cultivars is between 0.08 and 0.12 g/kg (Erbas *et al.*, 2005). For feeding purposes, the alkaloid content should be below 0.05% (Reinhard *et al.*, 2006). Alkaloid-free cultivars of white lupins are also available, and the development of these alkaloid-free mutants has allowed *L. albus* to be used as a protein source for animals. However, after growing in adverse conditions like drought stress, higher ambient temperatures, low soil pH, potassium deficiency or heavily weeded surroundigs, the alkaloid level of the so called "alkaloid free" varieties can be much higher and then cause restriction in feed intake (Weber and Krieg, 2024). Especially in *L. angustifolius* the alkaloid content varies considerably (table 2.6.1.1-3).

TABLE 2.6.1.1-3 Quinolizidine alkaloids (µg/g) in lupin seeds

Lupin variety	Alkaloid content	No. of samples
L. albus	143–226	5
L luteus	500–895	4
L. angustifolius	44–2120 median 138	18

REINHARD *ET AL.*, 2006

Particularly in poultry, but also in young piglets, the high level of NSP can be a limiting factor. Lupins have a negligible amount of starch, but high levels of soluble and insoluble NSPs and oligosaccharides (David *et al.*, 2024). These are partially fermented in the large intestine and cause wet litter, leading to footpad dermatitis, especially in young poultry.

2.6.1.2 Minerals and vitamins

The Ca content is comparable to soybeans and lower than in white-flowered peas (Table 2.6.1.2-1). The P concentration is comparable to other legumes, except for soybeans, which have a higher concentration. About 60% of the P is bound as phytate. Taking into account the wide variation between and within varieties, the other minerals are at about the same level as in peas and faba beans. The Cu content is low, especially compared to soybeans, which makes lupins an interesting feed for sheep. The level of iron is lower than in peas and only half that of soybeans. Manganese accumulation is a feature of *L. albus* leaves, especially in the low alkaloid lines. High levels of Mn in the diet can lead to secondary P and Fe deficiencies, as the absorption of both minerals is negatively affected by Mn. There is a negative correlation between Mn accumulation and seed yield, and selection for higher yield in white-flowered lupins could reduce seed manganese concentrations. Significant within-plant variation in Mn was also found in white lupins. The range of seed Mn concentrations in one plant was 1530 to 3750 mg/kg DM (Jezierny *et al.*, 2010).

The vitamin content is similar to that of cereals (Table 2.6.1.2-2). No vitamin A was found, but several studies reported the presence of carotenoids: β-carotene, lutein and zeaxanthin. However, the carotenoid content varied considerably between cultivars of the same lupin variety (Visvanathan *et al.*, 2020). The major tocopherols present in lupins are α-tocopherol, γ-tocopherol and δ-tocopherol, with γ-tocopherol as the most important isomer of vitamin E. These tocopherols are temperature sensitive. They are therefore destroyed by heat treatment such as toasting.

Lupin seeds are a good source of vitamin B. Vitamin C was only detected in some cultivars.

TABLE 2.6.1.2-1 Average mineral content (per kg DM) of lupin varieties

Item		Lupin variety		
		L. angustifolium	*L. albus*	*L. luteus*
Ca	g	2.8	2.6	2.6
P	g	4.7	5.5	5.5
DCP_{Pig}[1]	g	2.4	*	*
$DCP_{Poultry}$	g	*	5.0	5.0
Mg	g	1.9	2.0	2.0
Na	g	0.3	0.1	0.1
K	g	9.3	13	13
Cl	g	0.6	0.5	*
S	g	2.5	2.8	*
Co	mg	0.1	0.2	*
Cu	mg	5.5	4.5	*
Fe	mg	68	80	60
J	mg	0.2	*	*
Mn	mg	42	1927	79
Mo	mg	2.2	2.3	*
Zn	mg	53	30	53
Se	mg	0.1	0.1	*
DCAB[2]	meq	80	163	*
BE[3]	mmol/kg	237	338	*

1 Digestible P.
2 Dietary cation-anion balance.
3 Base excess.
* No data.

BELLOF AND FREITAG, 2021

2.6.2 *Feed value*

2.6.2.1 Protein

Protein quality for ruminants is determined by the N content, the amount of rumen undegraded protein (RUP) and the level of essential amino acids in the bypass protein. German feed tables indicate that lupins have a low to medium percentage of rumen undegraded protein (20% RUP) according to LfL Bayern (2017). In diets for high yielding dairy cows, protein sources with higher levels of bypass protein have to be added.

For monogastric animals, in addition to protein and AA content, digestibility values are an important feature of protein quality. The digestibility of lupins is higher

TABLE 2.6.1.2-2 Vitamin content of lupin varieties and cultivars (mg/100g DM)

Vitamin	*L. albus*				*L. angustifolius*	*L. luteus*
	cv. Multolupa[1]	cv. Marta[1]	Various cultivars[2]	Various cultivars[3]	Various cultivars[3]	cv. 44861
Vitamin A	*	*	*	0	*	*
α-Tocopherol	0.19	0.47	0	*	*	0.48
γ-Tocopherol	2.01	5.16	*	*	*	11.19
δ-Tocopherol	0.25	0.41	*	*	*	0.3.8
Thiamin (B1)	0.36	0.34	0.64	0.71	0.55	1.49
Riboflavin (B2)	0.61	0.65	0.22	0.24	0.33	0.85
Niacin	*	*	2.19	0.24	0.35	*
B6	*	*	*	0.4	*	*
B12	*	*	*	*	0	*
Pantothenic acid	*	*	*	0.83	0.16	*
Cholin	*	*	*	277.8	277.8	*
Biotin	*	*	*	100.0	100.0	*
Folic acid	*	*	*	0.05	0.05	*
Vitamin C	6.48	ND	4.8			ND

* No data.
ND Not detected.
1 Martinze-Villaluenga *et al.*, 2006.
2 Pereira *et al.*, 2022.
3 Bellof und Freitag, 2021.

than that of peas or faba beans and comparable to that of soybean meal (David *et al.*, 2024; Jezierny *et al.*, 2010; Table 2.6.2.1-1). This can partly be explained by the lower levels of neutral detergent fibre (NDF)-bound protein in lupins. These are only 29 mg/g CP, while peas contain 78 mg/g and faba beans 85 mg/g CP (Gdala, 1998). The reported crude protein content of sweet lupins ranges from 223 to 409 g/kg dry matter (David *et al.*, 2024). This variation is largely due to differences in cultivar, location and year of production, and agronomic management.

The AA profile is similar to other legume proteins, high in lysine and low in sulphur-containing AAs and tryptophan. These limiting AAs have to be supplemented in monogastric diets containing sweet lupins. The AA profile of *L. luteus* is very similar to that of soybeans of European origin.

TABLE 2.6.2.1-1 Amino acid (AA) content (g/kg DM) of lupin varieties (*L. angustifolius*, *L. luteus*, *L. albus*) and soybean (*Glycine max*) of European origin

Amino acid	*L. angustifolius*	*L. luteus*	*L. albus*	*Glycine max*
Lys	16	22	19	25
Met	2.0	2.6	2.9	5.5
Cys	5.0	8.2	6.3	5.7
Thr	12	14	14	15
Trp	2.7	3.3	2.7	5.5
Ile	16	17	18	18
Leu	25	31	28	30
Val	15	16	17	19

DURST *ET AL.*, 2021

In pigs, the digestibility of essential AAs for lupins is about 80% (*L. albus*) to 85% (*L. angustifolius* and *L. luteus*). In poultry, the digestibility is higher. *L. albus* has an average digestibility of 94%, compared with 85% for soybeans. The essential amino acid content of lupin species and soybeans is given in table 2.6.2.1-1 and their digestibility for pigs and poultry in Table 2.6.2.1-2. Both vary according to cultivar and growing conditions. Digestibility also depends on the age of the animals.

TABLE 2.6.2.1-2 Precaecal digestibility coefficient (%) for pigs and poultry of lupin varieties (*L. angustifolius*, *L. luteus*, *L. albus*) and soybean (*Glycine max*) of European origin

Amino acid	*L. angustifolius*		*L. luteus*		*L. albus*		*Glycine max*	
	Pig	Poultry	Pig	Poultry	Pig	Poultry	Pig	Poultry
Lys	87	87	87	87	82	92	87	88
Met	86	82	83	89	76	91	85	85
Cys	86	81	88	83	76	95	80	77
Thr	83	81	85	83	78	94	84	85
Trp	81	82	81	82	*	*	77	*
Ile	88	84	88	*	87	96	85	87
Leu	88	85	88	*	86	96	85	87
Val	84	82	84	82	74	94	84	86

* No data.

DURST *ET AL.*, 2021

2.6.2.2 Energy

Compared to wheat, the energy content of lupins differs between varieties. The higher crude fat content of *L. albus* (95 g/kg DM) compared to *L. angustifolius* (64 g/kg DM) and *L. luteus* (57 g/kg DM) is reflected in a generally higher ME content (Bellof and Freitag, 2021). Lupins have a low starch content, but comparatively high levels of NSPs and oligosaccharides. Due to the high fat and fermentable fibre content, the metabolisable and net energy of lupins for ruminants is higher than that of wheat. For pigs, the metabolisable energy is similar to that of wheat, but the net energy is lower due to the fibre content. Metabolisable energy for poultry and horses is lower than for wheat. Dehulling the seeds increased the concentration of nutrients and the energy value for chickens. The oligosaccharide content seems to limit its wider use in poultry diets (David *et al.*, 2024).

As with other plants, the energy content of lupins can vary in response to genetic and environmental variation.

2.6.3 *Feeding recommendations*

Legumes in general are a valuable source of protein and are feed material for most domestic animals. However, due to their alkaloid and fibre contents, acceptable levels of lupins in the diet vary depending on the lupin variety and cultivar, as well as the species and age of the animal. Several feeding trials have been carried out over the last few decades to determine the acceptable levels of modern lupin cultivars.

However, when using lupins in compound feed, it must be kept in mind that their seeds are larger and therefore more difficult to dry than cereal grains. Residual moisture above 14%, for example due to overgrowth or uneven ripening, can cause mould growth. Animals reject these lupins. Besides alkaloids, mould can be a reason for reduced feed intake in diets with lupins.

2.6.3.1 Cattle and small ruminants

To assess the nutritional value of plants for ruminants, it is important to consider their effects on rumination and on rumen fermentation, in addition to their nutritional value. The stimulation of chewing and rumination depends on particle size and mechanical resistance to chewing. The latter is estimated by the level of structurally active fibres (NDF, ADF, CF) and their chemical composition (Table 2.6.3.1-1). The fibre content of lupins is much higher than that of wheat, especially the ADF content. However, as lupins are milled for feeding, they do not promote rumination.

In ruminant feeding, protein quality is mainly determined by the nitrogen content and the rate and extent of ruminal protein degradation. Amino acid composition is only considered relevant in ruminal bypass protein. The ruminal fermentation rate depends on the level of feed intake, which determines the length of time the feed remains in the rumen. The structure of the protein and its incrustation with poorly fermentable fibre also determine the extent of protein conversion in the rumen. A certain amount of protein passes through the rumen unchanged. This is reflected in the undegraded protein (RUP) value, which is low for all lupin varieties and may limit their inclusion in diets for high yielding cows. Several trials have been carried out to increase the RUP of lupins with the aim of replacing soybeans in the diet.

In a feeding trial with dairy cows, Pries *et al.* (2007) tested the effects of hydrothermal treatment (135 °C, Lupitherm® treatment) on *L. angustifolius*. Heat treatment resulted in a 3.5-fold increase in RUP content. Chemical analysis revealed a shift in the protein fractions from B1 to B2, while the rapidly degradable fraction A1 and the poorly degradable fractions B3 and C were not affected. Cows fed the treated *L. angustifolius* (1.4 kg/d; 46% of the concentrates) had significantly higher milk and milk protein yields.

Pieper *et al.* (2007) tested the effects of pressure heat treatment (> 130 °C at 30 bar; Opticon®, Deutsche Tiernahrung GmbH & Co. KG) on *L. angustifolius*. A feeding trial with rats showed that the treatment did not affect the ileal protein digestibility. Chemical analysis of the protein showed a shift from A1 and B1 to B2 fractions, while B3 and C were unaffected. RUP was twice that of untreated lupins (UDP 8%/h: 23.6 versus 11.4%). Milk yield in cows fed the treated lupins at 2.9 kg DM/d was 8% higher than in the group fed the untreated lupins (35.2 versus 32.6 kg/d). Milk protein and milk fat percentages did not change. In a second feeding trial, an expander treatment on a mixed lupin/rapeseed feed component was compared with an untreated soybean/rapeseed combination. The expander-treated diet had a higher B2 fraction (70.4 versus 58.9%) and a lower B3 fraction (5.9 versus 19.1%). RUP at 8%/hr was also lower (45.6 versus 56.0%). At the beginning of the trial, there was no difference in milk yield between the two groups. However, after 45 days the milk yield of the soybean/rapeseed group decreased, while that of the lupin/rapeseed group remained stable at 35 kg/d. Milk protein and milk fat content did not differ between the groups. All three feeding trials show the positive effects of heat treatment of lupins on protein efficiency in dairy cows. This is of particular interest in high yielding production systems that are trying to avoid feeding soybeans.

However, a feeding trial with high amounts of heat treated *L. angustifolius* revealed some negative effects compared to diets with a combination of rapeseed meal /*L. angustifolius* or rapeseed/soybean meal (Kluth *et al.*, 2007). Diets contained either up to 4 kg heat treated lupins, 2.5 kg lupin plus rapeseed or no lupins and rapeseed/soybean meal instead. Daily feed intake was significantly lower in the lupin group than in the two other groups. RUP was slightly higher in heat treated lupins compared to soybean meal (37 to 31%), but much lower than in rapeseed (58%; table 2.6.3.1-1) and milk yield was significantly lower in the pure lupin group than in the two other groups. Milk protein content was highest in the rapeseed/soybean meal group, as was ruminal protein degradation and microbial protein synthesis (195 versus 153 g/kg OM; table 2.6.3.1-2). Very high amounts of *L. angustifolius* may reduce feed intake and the efficiency of ruminal protein synthesis. Moreover, if lupins contain high levels of alkaloids, these may be transferred into the milk and may thus be potentially toxic for humans (Engel *et al.*, 2022).

TABLE 2.6.3.1-1 Characteristics of heat treated *L. angustifolius* and untreated rapeseed meal (RES) and soybean meal (SES)

Parameter		*L. angustifolius*	RES	SES
CP	g/kg	363	384	454
NDF	g/kg	262	320	195
ADF	g/kg	211	227	135
DCOM[1]	%	93	76	90
DCNDF[2]	%	97	60	93
RUP of XP	%	37	58	31
ME	MJ/kg DM	14.4	11.6	13.5
NEL	MJ/kg DM	9.1	7.0	8.4

1 Digestibility coefficient of organic matter.
2 Digestibility coefficient of NDF.
KLUTH *ET AL.*, 2007

TABLE 2.6.3.1-2 Efficiency of microbial protein synthesis (MP) of a total mixed ration (TMR) with heat treated *L. angustifolius* (L.A.), *L. angustifolius* and rapeseed meal (L.A./RES) and rapeseed meal-soybean meal (RES/SES)

Parameter		TMR composition		
		L.A.	L.A./RES	RES/SES
L.A.	g/kg TMR	18.8	11.0	0
MP	g/kg OM	153[a]	162[a,b]	195[b]
MP	g/MJ ME	7.9[a]	8.3[a]	10.1[b]

a,b $p < 0.05$.
KLUTH *ET AL.*, 2007

Replacing soybean meal with lupin seeds has also been tried in ruminant fattening. Sami *et al.* (2010) fed Simmental beef calves from 276 to 670 kg bodyweight with either *L. angustifolius* or soybean meal as the main protein source. They found no differences in feed intake, feed conversion rate, daily weight gain or carcass and meat quality traits between the two groups and concluded that sweet lupins can replace soybean meal in beef production.

TABLE 2.6.3.1-3 Recommendations for the inclusion of sweet lupins in cattle feed (all lupin varieties)

Cattle group		g/100 kg BW[1]	kg/animal/day	% of compound feed
Calves (Up to 4 months of age)		*	*	20
Heifers (From 4 months on)		400	*	40
Dairy cows	Raw lupin	500	3.5	35
	Heat treated	500	4.0	40
Suckler cows		500	1.5 (300 kg BW)	*
			2.5 (500 kg BW)	*
			3.5 (700 kg BW)	*
Beef cattle		500	2.0	45

1 BW: body weight.
* No recommendation.
HOFFMANN AND STEINHÖFEL, 2023

In a feeding trial with Romanov × Suffolk lambs, soybean meal was replaced by *L. albus* with a low alkaloid content (0.03%). No differences were found in dry matter intake, average daily gain and feed conversion ratio. Stanford *et al.* (1996) concluded that soybean meal can be replaced by lupins, provided that the alkaloid content is low.

The use of either lupins and/or peas as the main protein source has also been tested in lamb fattening (Lestingi *et al.*, 2016). The protein source had no effect on carcass weight, average daily gain and feed conversion ratio. Meat characteristics, like chemical composition, pH, colour and tenderness were also unaffected by the protein source. However, meat from the lupin group contained higher levels of saturated fatty acids. The authors concluded that lupin as a sole protein source did not affect carcass traits, but had a negative effect on intramuscular fatty acid composition, while the use of peas and lupins together eliminated the negative effects of lupins. Young sheep accept up to 20% of sweet lupins in compound feed, independent of the lupin variety (table 2.6.3.1-4). For mature sheep, a maximum of 290 g of sweet lupins per 100 kg BW is recommended. For dairy goats and their offspring, the inclusion in the diet should be less (table 2.6.3.1-4).

TABLE 2.6.3.1-4 Recommendations for the inclusion of sweet lupins of unknown cultivar in feed for small ruminants

Animal group	g/100 kg BW[1]	g/animal/day	% of concentrates
Sheep	290	200	*
Lambs			
Sheep	*	*	25
Goats	*	*	15
Dairy goats	170	100	*

1 Body weight.
* No recommendation.
HOFFMANN AND STEINHÖFEL, 2023

2.6.3.2 Pigs

Feed quality for pigs is mainly determined by the energy content and the level of digestible essential amino acids. Amino acid availability of lupins, tested in cannulated pigs, was high and intervarietal differences were small. Therefore, lupins are a good source of protein in pig diets.

The energy content depends on the digestibility of nutrients in the small intestine. In pigs, increasing the level of lupins and reducing the amount of cereals in the diet did not change its digestible energy content, but the proportion of energy digested at the end of the small intestine decreased significantly due to the high levels of NSP and α-galactosides present in lupins. This reduces the net energy content compared to cereals. Lupins are high in microbial degradable fibre. About 40% of the digestible energy in lupins is known to come from microbial fermentation in the large intestine. However, fibre composition and content varies between lupin cultivars. This explains the wide range of net energy found in *L. albus* (3.23 to 10.89 MJ/kg DM) and *L. angustifolius* (6.15 to 10.54 MJ/kg DM) (Jezierny *et al.*, 2010).

The main limiting factor for the inclusion of lupins in pig diets is the alkaloid level. The tolerable level of lupins in pig diets depends on both the total amount of dietary alkaloids and their composition, which varies between lupin varieties and cultivars. Young piglets are more sensitive to the bitter taste than older pigs. For example, according to Buraczewska *et al.* (1993), pigs do not tolerate more than 0.12 g/kg of alkaloids from *L. albus*, whereas diets containing up to 0.45 g/kg of alkaloids from *L. luteus* have been consumed without any negative effect on feed intake. Differences in alkaloid composition between lupin species may explain, at least in part, the variation in feed intake response. However, observations regarding the level of lupins of the same cultivar in feeding trials are controversial. Inclusion of 120 g/kg of the *L. albus* cultivar *Buttercup* with an alkaloid content greater than 0.5 g/kg DM significantly reduced feed intake in weaned piglets by up to 21%, but the same cultivar with an alkaloid content of 0.1 g/kg DM had no effect on feed intake when included in the diet at the same level.

Therefore, alkaloid composition and content seem to play an important role in the feeding of lupins.

Gdala *et al.* (1996) did not find any growth retardation in young pigs fed diets supplemented with *L. angustifolius* (410 g/kg diet) or *L. luteus* (320 g/kg diet) compared with diets based on barley and soybeans. Yet, lower feed efficiency was observed in piglets fed a diet containing 375 g/kg *L. albus* seeds with an alkaloid content of 0.23 g/kg DM (Jezierny *et al.*, 2010). With lupins of unknown cultivar, only up to 150 g/kg of *L. albus* should be included in diets for growing pigs (30–60 kg body weight). The recommended inclusion level for lupins in diets for finishing pigs is 200 g/kg. However, if they grew under adverse conditions, even 100 g/kg feed can reduce feed intake (Weber and Krieg, 2024).

The total alkaloid content in diets for growing and finishing pigs fed ad libitum should not exceed 0.2 and 0.33 g/kg respectively. As the alkaloid level of current sweet lupin cultivars is usually quite low, these recommended limits will normally not be exceeded when modern lupin cultivars are used as a component of commercial pig diets. However, alkaloid levels in lupins can vary depending on growing conditions such as drought stress or phosphorus-potassium fertilisation (Jezierny *et al.*, 2010). It is therefore necessary to monitor alkaloid levels in lupins, especially if they are used in commercial diets.

TABLE 2.6.3.2-1 Recommendations for feeding lupins of unknown cultivar to pigs (% of diet)

Animal group	Body weight / reproductive cycle	Lupins % of diet	
		Raw	Toasted
Piglets	< 15 kg	0	5
	> 15 kg	5	15
Fattening pigs	30–60 kg	15	15
	60–120 kg	20	20
Sows	Pregnant	25	25
	Lactating	20	20

HOFFMANN AND STEINHÖFEL, 2023

2.6.3.3 Poultry

As long as essential AAs are balanced in the diet, the inclusion of lupins in poultry diets is mainly limited by their carbohydrate profile and alkaloid content. Soluble oligosaccharides are fermented in the gut and can lead to wet litter. The content of these carbohydrates limits the wider acceptance of sweet lupins in poultry diets (David *et al.*, 2024).

However, the results of feeding trials on the level of lupins that can be included in broiler diets are controversial. Farrell *et al.* (1999) tested different inclusion levels of lupins and suggested an inclusion level of less than 100 g/kg for broilers. Feeding trials by Nalle *et al.* (2012) showed that *L. angustifolius* can be included at 200 g/kg in broiler starter diets when the diets are properly balanced for AME and digestible AAs. Van Barneveld (1999) indicated that lupins could be used up to 250 g/kg in broiler diets. According to Hughes *et al.* (1998), whole lupin seeds can be used up to 200 and 300 g/kg in wheat- and maize-based broiler diets, respectively, but wet litter problems may occur at high lupin concentrations. Perez-Escamilla and Vohra (1987) also found that a lupin inclusion level of 300 g/kg could support broiler performance without adverse effects. The observed discrepancies may be explained by cultivar differences in the content of alkaloids and NSPs and the failure to balance AME in feed formulations (David *et al.*, 2024).

The nutritional value of lupins is largely determined by the composition and concentration of alkaloids in the seed. As discussed above, these bitter substances can affect feed intake and growth in poultry and limit the use of lupins. Varieties with low alkaloid content are well accepted. Feeding 400 g/kg of *L. albus* (alkaloid content < 0.1 g/kg; cv Hanti) to broilers up to 8 weeks of age showed no adverse effects on growth, feed efficiency or carcass characteristics (Olver and Jonker, 1997). In contrast, Olkowski *et al.* (2005) reported a significant reduction in feed intake and weight gain in broilers fed the same amount also of white lupins. These differences are most likely due to differences in alkaloid concentrations and their composition in different *L. albus* cultivars.

Sweet lupins can also be included in laying hen diets to replace other protein sources. Whatkins and Mirosh (1987) reported no adverse effects on laying characteristics, only increased egg yolk colour in hens fed diets containing 100–300 g/kg *L. albus* with low alkaloid content (cv Ultra). Perez-Maldonado *et al.* (1999) also found no adverse effects of feeding *L. angustifolius* at 250 g/kg on laying hen performance compared with field peas and faba beans. However, digesta viscosity and pancreas weight were increased. This may be due to the NSP content of lupins.

2.6.3.4 Horses

Horses with higher energy and/or protein requirements (growing, lactating, heavy work) are fed concentrates. The main ingredients of these concentrates are oats and soybean meal, the main nutrients starch and protein. To replace part of the

TABLE 2.6.3.3-1 Recommended inclusion of sweet lupins of unknown variety and cultivar in poultry feed (% of diet)

Poultry group	Sweet lupins Raw	Sweet lupins Toasted
Chicken	15	5
Pullets	20	15
Laying hens		
Brooder	30	15
Layer	20	20
Broilers	15	15
Turkey		
Breeding	10	10
Starter (P1/P2)[1]	15	15
Grower (P3–4)	15	20
Finisher (P5–7)	15	20
Ducks, Geese		
Breeding	15	15
Fattening	20	20

1 Production period.
HOFFMANN AND STEINHÖFEL, 2023

soybean meal, sweet lupin seeds with a low alkaloid content are a suitable source of protein and amino acids. However, the high content of fermentable oligosaccharides, particularly raffinose, may cause gas formation in the large intestine (Coenen and Vervuert, 2020).

Lupin seeds vary in size and therefore need to be carefully dried after harvest. Seeds with a moisture content above 14% may become infected with moulds that produce mycotoxins. These toxins are most likely responsible for reported organ damage in horses such as liver damage, emaciation, jaundice or central nervous system disorders after lupin feeding. Sweet lupins with an alkaloid content of less than 0.01% and in good hygienic condition can be fed up to 0.2 kg/100 kg BW per day without further treatment (Coenen and Vervuert, 2020).

To assess the digestibility of lupin seed, the inclusion of up to 14% *L. albus* in a crip feed was tested. The apparent digestibility of organic matter, nitrogenous components and fibre were 57.6%, 63.0% and 63.6%, respectively. The low digestibility of the diets was most likely due to antinutritional factors in the lupin seed (Šajdler and Zeman, 2005).

2.6.3.5 Fishes and shrimps

Sinem Zeytin, Monika Weiß and Vanessa Fuchs

Sweet lupins, special cultivated forms of the white lupin (*Lupinus albus*), the yellow lupin (*L. luteus*) and the blue lupin (*L. angustifolius*) are increasingly used as a sustainable alternative protein source (ranging from 30% to 40% CP) (Petterson and Mackintosh, 1994; Glencross *et al.*, 2007) to replace fishmeal and soybean meal in fish and shrimp feeds (Szczepański *et al.*, 2022; Van de Noort, 2024). These lupins have been studied to assess their nutritional value and biological effect on different important farmed aquaculture species (Table 2.6.3.5-1). Besides being a valuable source of protein, lupin seeds also have a balanced amino acid profile although it is relatively low in methionine, lysine and threonine compared to fish meal (Gatlin *et al.*, 2007). These AAs are critical for fish growth. Therefore, in order to meet the specific proportion of essential AAs requirements of different aquaculture species, supplementations were considered for individual cases (Table 2.6.3.5-1). Lupins also contain dietary fibre and non-starch polysaccharides which can affect digestibility and energy utilization in fish. However, it was determined that processing methods such as dehulling, extrusion, fermentation, thermal treatment and enzyme treatment (Petterson and Mackintosh, 1994; Burel *et al.*, 2000; Glencross *et al.*, 2003a; Glencross *et al.*, 2004; Mieczkowska *et al.*, 2005; Glencross *et al.*, 2008, 2012a,b) can help to reduce the fibre content and antinutritive factors like alkaloids, oligosaccharides, and phytic acid in the seed and improve nutrient availability. All these technological advancements as well as being more sustainable and cost-effective compared to traditional protein sources, especially when produced locally, make lupin an increasingly viable option for fish farmers and aquafeed producers.

Research is ongoing to optimize lupin inclusion levels in the diets of different fish species and their life stage to ensure maximum growth performance and health benefits. Over the last decades, lupins have proved their potential as feed ingredients, especially for rainbow trout (Glencross *et al.*, 2004; Gatlin *et al.*, 2007; Glencross *et al.*, 2007; 2012a,b), and Atlantic salmon (Carter and Hauler, 2000; Refstie *et al.*, 2000; Glencross *et al.*, 2003b) which need high levels of protein in their diet. Studies have shown mixed results regarding growth performance when lupin meal replaces fishmeal in rainbow trout diets. Moderate inclusion levels up to 20–30% have been generally well tolerated without significant adverse effects on growth and feed efficiency. However, when targeting protein sources alternative to fishmeal and/or soybean meal, it is not only important to focus on the protein source itself, but also on the health status and fish welfare.

Feeding experiments with different inclusion levels of lupin meal (*L. albus*) fed to rainbow trout (*Oncorhynchus mykiss*) juveniles demonstrated that lupin meal can be included in diets as a replacement of fishmeal up to 30% without any malnutrition effect on growth performance and health (Acar *et al.*, 2018). Similar results were achieved by Farhangi and Carter (2001), where dehulled lupins (*L. angustifolius*) were included at up to 40% of a rainbow trout diet without a significant effect on growth performance and nutrient utilization. However, higher inclusion levels can affect the digestive system and gastrointestinal tract by up to 15% inclusion level. On the other hand, at a maximum inclusion level of 50% of whole grain lupin meal (*L. albus*), when lysine and methionine AAs were supplemented in extruded diets for rainbow trout, no significant trends were observed on growth, feed utilization, apparent digestibility coefficients and whole-body composition (Borquez *et al.*, 2011). In addition, with increasing inclusion level of lupin, the fish fillet fatty acid concentrations showed an increase in polyunsaturated omega-3 fatty acid and a decrease in eicosapentaenoic acid (EPA). However, the concentration of docosahexaenoic acid (DHA), which is beneficial for both fish health and human nutrition, remained stable in the muscle. In another study, Glencross *et al.* (2005) showed that *L. luteus* and *L. angustifolius* in the form of lupin protein concentrates were highly palatable and provided equivalent nutritional value to the rainbow trout in two inclusion levels, 20% and 40%.

Digestibility of nutrients from lupin meal can be lower compared to fishmeal, primarily due to its fibre content and the presence of anti-nutritional factors, but digestibility it is still generally comparable to fishmeal. Recently, inclusion of 20% Andean lupin (*L. mutabilis*) seed meal showed no adverse effects on growth performance and nutrient digestibility on rainbow trout juveniles (Serrano *et al.*, 2024). In addition, extruded lupin seed meal has been found to offer good digestibility coefficients especially for protein, and energy by trout and turbot (*Psetta maxima*). An inclusion of 50% of extruded lupin in turbot diets supported growth performance when compared to no inclusion in a control diet formulation (Burel *et al.*, 2000).

Several studies examined the digestibility of regional lupin proteins in the diet for Atlantic salmon (*Salmo salar*) and suggested a dietary inclusion of lupin proteins at a level between 20% and 30% (Szczepański *et al.*, 2022) without negative impact on growth performance. For example, lupin digestibility has been shown to exceed that of soya in Atlantic salmon (Glencross *et al.*, 2004). Replacing fishmeal with raw or fermented lupin with *Lactobacillus plantarum* did not affect the growth, apparent digestibility coefficient and immune status of juvenile Atlantic salmon. It even improved performance when supplemented at 15%, whereas a 30% replacement of fishmeal with fermented lupin showed no improvement (Rodríguez-Estrada *et al.*, 2020). The digestive function and intestinal integrity are crucial aspects to consider when using plant proteins in aquafeeds for species like Atlantic salmon. Carter and Hauler (2000) evaluated the nutritional value of diets containing a lupin protein concentrate and compared it with defatted soybean meal and a pea protein concentrate. The three protein sources were included in diets at either 25% or 33%, replacing fishmeal protein in the diet. The highest digestibility was observed in the diets in which the *L. angustifolius* protein concentrate replaced 33% of the fishmeal. No lupin seed meals or protein concentrates altered the morphology of the salmon intestine as did the soybean meal. The protein in the lupin protein concentrates also appeared more available than the protein in the lupin seed meals (Refstie *et al.*, 2006). Previous experiments using cold-pressed pellets for Atlantic salmon suggested that 40% fishmeal protein replacement with pea or lupin protein concentrates was feasible (Carter, 2000).

In addition to salmonids, lupin meal was tested in diets for common carp, barramundi, tilapia and shrimps. Most of these studies have shown that lupins may replace fishmeal even at 40–50%, which is a highly promising result (Szczepański *et al.*, 2022).

High inclusion rates up to 50% of lupin seed meal can be used as a replacement for fishmeal in diets for the carnivorous European Seabass (*Dicentrarchus labrax*), without adverse effects on growth performance. In addition, no significant differences were observed between fermented and unfermented lupin seed meal groups (Zeytin *et al.*, 2018).

Protein digestibility coefficients for gilthead seabream (*Sparus aurata*) fed lupin seed meal diets were similar to the control group and 10% higher than those for the soybean meal diets. A significant reduction in trypsin activity was observed in fish fed the lupin seed meal diets, and for soybean meal diets when the substitution level reached 30%. However, the use of up to 20% treated lupin meal showed no significant differences in final body weight, feed efficiency, protein efficiency and protein production value compared to the reference diet (Robaina *et al.*, 1995). Similar results obtained by Pereira, and Oliva-Teles (2004) demonstrated that lupin seed meal can replace up to 30% of fishmeal protein in diets for gilthead sea bream juveniles with no negative effects on growth performance. Furthermore, micronization of lupin seeds improved its dietary value and promoted significantly higher growth rates than raw lupin seeds. The gut microbiota of gilthead seabream was affected during a feeding trial in which the fish were fed a control diet with fishmeal as the sole protein source and a plant protein source introduced at the level of 20% of lupin meal (Silva *et al.*, 2011).

Another study showed that partial, but not total, replacement of soybean protein with lupin seed protein in juvenile tilapia (*Oreochromis niloticus*) diets resulted in better, or at least equal, growth and feed performance. Dehulling further enhanced the growth and performance of lupin seed meal. Alkaloid removal improved feed performance but not the growth (Chien and Chiu, 2003). More recently Bowyer *et al.* (2020) have shown positive results on gut health when a solid-state fermentation product, Synergen™, was supplemented to diets containing both blue and yellow-flowered lupin meals for tilapia.

For juvenile barramundi (*Lates calcarifer*) lupin fermented by *Lactobacilli* improved the lupin's nutritional quality (significantly reduced the amount of antinutrients and improved the amino acid profile), allowing higher inclusion level in barramundi diets (Van Vo *et al.*, 2015). The growth rates of the juvenile barramundi were not adversely affected by the different inclusion levels of fermented lupin at 30%, 45%, 60%, and 75% in the diets. There was even a significant increase in the final weight and length of the barramundi fed 45% and 60% fishmeal replacement diets. Survival rates were greater than 93% for all diets. Feed conversion rates were unchanged between the diets, with the exception of significantly higher conversion rates in the 75% fishmeal replacement diet.

A recent study from Anwar *et al.* (2020) showed that white lupin meal can be included by up to 25% in Common carp (*Cyprinus carpio*) diets, without showing adverse effects on body composition and physiological status. In addition, solid state fermentation of lupin meal even enhanced growth performance, nutrient utilization (feed conversion rate, protein efficiency ratio), gut and liver morphology by replacing soy protein concentrate in carp diets.

Diets containing lupin seed meal (*L. angustifolius*) fed to whiteleg shrimp (*Litopanaeus vannamei*) as fishmeal replacement showed positive results at levels of up to 10% (–20%) of the total feed (= 30–40% of animal protein) with no growth or metabolic impairment compared to a commercial shrimp feed. A moderate lupin inclusion functioned as bioactive compound stimulating the shrimp's immune system as well (Weiss *et al.*, 2020). However, inclusion levels greater than 25% resulted in a progressive decline in shrimp performance. However, when the diets contain marine ingredients at inclusions of at least 50%, specific feeding effectors do not appear to be necessary to maintain the palatability of the shrimp diets (Smith *et al.*, 2000). On the other hand, the protein and amino acids in lupin seed meal and in the protein concentrate are well digested by shrimp, with an average apparent crude protein digestibility of > 90% (Smith *et al.*, 2006). The evaluation of the Andean lupin (*L. mutabilis*) revealed that at least half of the fishmeal protein in the diet for *L. vannamei* can be replaced without negatively influencing growth and feed conversion rate (Molina-Poveda *et al.*, 2013).

A feeding experiment with juvenile black tiger shrimp (*Penaeus monodon*) investigated new lupin cultivars that would affect the nutritional value of the seed meal. This study has demonstrated that lupin seed meal can be used to replace at least 40% of the fishmeal protein in diets for *P. monodon*, and that the new cultivars perform equally to solvent-extracted soybean meal when used on a protein-equivalent basis (Smith *et al.*, 2007). Another dose-experiment on replacement of defatted soybean meal by dehulled lupin meal (*L. albus*) showed no statistically significant differences in weight gain, feed intake, feed conversion rate, protein conversion efficiency and apparent protein utilization of juvenile *P. monodon* fed diets with 25 and 50% replacement (Sudaryono *et al.*, 1999). In a review from Rajeev and Bavitha (2015), it is stated that results of a feeding preference test also indicated that *L. angustifolius* diet was more attractive for juvenile *P. monodon* than *L. albus* diet. In addition, *L. albus* and the nutritive value of *L. angustifolius* diet was equal to that of soybean control diet by *P. monodon*.

A comprehensive study about the effect of different lupin species, cultivars and inclusion levels on the physical properties of extruded feed pellets was conducted by Glencross *et al.* (2010). Including increasing inclusion rates at 10, 20 and 30% of lupin in extruded diets led to a significant increase in bulk density, sink rate, in crush and shear force as well as in the water retention capacity. These results are directly linked to the carbohydrate and especially the non-starch polysaccharide levels present in the different species and cultivars tested, which also has been described by Cheung (1990). The inclusion of lupin seed meal in diets can be used to increase bulk density and settling time, both of which increase with increasing lupin content. Pellet hardness was also affected by legume inclusion, showing highest effect with *L. angustifolius*, but varying effects among the different cultivars. Higher pellet hardness improves pellet durability, which makes a feed more suitable for automated feeders and facilitates cleaning in land-based aquaculture systems (e.g. recirculating aquaculture systems), and may also improve nutrient utilization in the fish gut in some species (Baeverfjord *et al.*, 2006). Conversely, high pellet hardness may also hinder nutrient uptake (Samuelsen *et al.*, 2021) and prolong gastric emptying rate and reduced feed intake in Atlantic salmon. So an optimized balance between pellet hardness, gastrointestinal transit time and nutrient absorbability needs to be considered (Bogevik *et al.*, 2021).

L. angustifolius was found to have the highest water holding capacity, which on the one hand reduces production costs due to lower friction in the barrel with higher fluidity, thus reducing the depreciation of the extruder, which is reported to be the highest cost in feed extrusion. On the other hand, the higher water holding capacity allows extrusion at lower temperatures with higher starch gelatinization and less risk of protein damage.

TABLE 2.6.3.5-1 Executive summary of studies testing sweet lupins ingredients in aquafeeds for various species

Species	Initial mean weight (g)	Lupin source	Product/ process	Replaced source	Inclusion rate (%)	Recommendable inclusion rate in %	Notes	Reference
Rainbow trout (*Oncorhynchus mykiss*) (*Salmo gairdneri*)*	45	*L. albus*	Thermal treatment	FM[a]	10, 20, 30, 40	30	Thermal heating did not improve nutritinal quality	De La Higuera *et al.*, 1988
	14	*L. albus*	Dehulled	FM	15, 30, 45, 60	Up to 30		Acar *et al.*, 2018
	47	*L. angustifolius*	Dehulled	FM	10, 20, 30, 40, 50	Up to 40	Rainbow trout utilize dehulled lupin protein as efficiently as fishmeal protein but have a lower ability to utilize the energy content of dehulled lupin at higher inclusion levels	Farhangi & Carter, 2001

a Fishmeal.

TABLE 2.6.3.5-1 Executive summary of studies testing sweet lupins ingredients in aquafeeds for various species (*cont.*)

Species	Initial mean weight (g)	Lupin source	Product/ process	Replaced source	Inclusion rate (%)	Recommendable inclusion rate in %	Notes	Reference
	54	*L. albus*	Whole grain-extruded	FM	30, 40, 50	Up to 5	(+) methionine and lysine	Borquez *et al.*, 2011
	37	*L. albus*	Extruded	FM	15, 25, 35	25	(+) methionine and lysine	Hernández *et al.*, 2013
	Experiment I: 23 Experiment II: 27 Experiment III: 34	*L. albus*	Extruded	FM	30, 50, 70	50	Higher fat deposition in fish was observed	Burel *et al.*, 1998
	100	*L. albus*	Extruded	FM	30	30		Burel *et al.*, 2000
	66	*L. albus*		FM and wheat	10, 15, 20	Up to 20		Borquez *et al.*, 2011
	330	*L. luteus*	As additive, extruded		0.05, 0.075, 0.1, 0.25, 0.5, 1, 5	0.1		Serrano, 2024

TABLE 2.6.3.5-1 Executive summary of studies testing sweet lupins ingredients in aquafeeds for various species (*cont.*)

Species	Initial mean weight (g)	Lupin source	Product/ process	Replaced source	Inclusion rate (%)	Recommendable inclusion rate in %	Notes	Reference
	266	*L. angustifolius*	Protein concentrate	FM	30	30%		Glencross *et al.*, 2004
	36	*L. angustifolius* & *L. luteus*	Protein concentrate	FM	10, 20, 30, 40	20, 40	(+) methionine and lysine	Glencross *et al.*, 2006
	30	*L. angustifolius* & *L. luteus*	Extruded	FM	10, 20, 30, 40	40	(+) methionine and lysine	Glencross *et al.*, 2011
Atlantic salmon (*Salmo salar*)	47	*L. angustifolius*	Protein concentrate, dehulled, milled, extruded	FM	22, 29	22	(+) methionine	Carter and Hauler, 2000
	66	*L. angustifolius*	Protein concentrate	FM	30	30		Glencross *et al.*, 2004
	90	*L. albus* & *L. angustifolius*		FM	25	25		Salini and Adams, 2014

TABLE 2.6.3.5-1 Executive summary of studies testing sweet lupins ingredients in aquafeeds for various species (*cont.*)

Species	Initial mean weight (g)	Lupin source	Product/ process	Replaced source	Inclusion rate (%)	Recommendable inclusion rate in %	Notes	Reference
	176	*L. angustifolius* & *L. luteus*	Protein concentrate, extruded	FM	30	30		Refstie, 2006
	3.5	*L. albus*	(Raw & fermented with *L. plantarum* 299v, (extruded))	FM	15, 30	15	No different effects on fish with fermented or non-fermented diet	Rodríguez-Estrada *et al.*, 2020
Gilthead seabream (*Sparus aurata*)	Experiment I: 40 Experiment II: 72	*L. angustifolius*		FM	10, 20, 30	20	Must not exceed 20% of total protein to prevent lipid deposition in the liver.	Robaina *et al.*, 1995
	19			FM	20	20	(+) methionine, lysine, leucine	Silva *et al.*, 2010, 2011
	42	*L. angustifolius*	Infrared radiation – micronized		10, 20, 30	30	Micronization of lupin seeds improves its dietary value	Pereira & Oliva-Teles *et al.*, 2004

TABLE 2.6.3.5-1 Executive summary of studies testing sweet lupins ingredients in aquafeeds for various species (*cont.*)

Species	Initial mean weight (g)	Lupin source	Product/ process	Replaced source	Inclusion rate (%)	Recommendable inclusion rate in %	Notes	Reference
European seabass (*Dicentrarchus labrax*)	57	*L. angustifolius*	Extruded, fermented	FM	25, 50, 75, 100	50		Zeytin *et al.*, 2018
Barramundi (*Lates calcarifer*)	6,9	*L. angustifolius*	Fermented		30, 45, 60, 75	30, 45, 60		Van Vo, 2015
Turbot (*Psetta maxima*)	110	*L.albus*	Extruded	FM	30	30		Burel *et al.*, 2000
Nile tilapia (*Oreochromis niloticus*)	0.3	*L. angustifolius*	Whole, dehulled	Soymeal	15, 30, 50	15, 30	Dehulled lupin seed meal diets resulted in better growth and feed performance than the whole LSM diets. Alkaloid removal improved feed performance but not the growth	Chien and Chiu, 2003

TABLE 2.6.3.5-1 Executive summary of studies testing sweet lupins ingredients in aquafeeds for various species (*cont.*)

Species	Initial mean weight (g)	Lupin source	Product/ process	Replaced source	Inclusion rate (%)	Recommendable inclusion rate in %	Notes	Reference
	36	*L. albus* & *L. angustifolius*	Dehulled		15	15	(++) solid state fermentation	Bowyer *et al.*, 2020
Common carp (Cyprinus carpio)	16	*L. albus*		Soymeal	12.5, 25	25	(++) solid state fermentation	Anwar *et al.*, 2020
Black tiger shrimp	3.3	*L. angustifolius*		FM	40	40		Smith *et al.*, 2007
(*Penaeus monodon*)	4	*L. albus*	Dehulled	Defatted soymeal	25, 50, 75, 100	50		Sudaryano *et al.*, 1999
Whiteleg shrimp (*Litopenaeus vannamei*)	1.23	*L. mutabilis*			0, 25, 50, 75 and 100			Molina-Poveda *et al.*, 2013
	3.6	*L. angustifolius*			10, 20, 30	10, 20		Weiss *et al.*, 2019

References

Acar Ü., Kesbiç O.S., Yilmaz S., Karabayir A., 2018: Growth performance, haematological and serum biochemical profiles in rainbow trout (*Oncorhynchus mykiss*) fed diets with varying levels of lupin (*Lupinus albus*) meal. Aquaculture research, 49, 2579–2586.

Anwar A., Wan A.H., Omar S., El-Haroun E., Davies S.J., 2020: The potential of a solid-state fermentation supplement to augment white lupin (*Lupinus albus*) meal incorporation in diets for farmed common carp (*Cyprinus carpio*). Aquaculture reports, 17, 100348.

Baeverfjord G., Refstie S., Krogedal P., Åsgård T., 2006: Low feed pellet water stability and fluctuating water salinity cause separation and accumulation of dietary oil in the stomach of rainbow trout (*Oncorhynchus mykiss*). Aquaculture, 261, 1335–1345.

Bellof G., Freitag M., 2021: Körnerleguminosen. Editors: Durst L., Freitag M., Bellof G., 2021: Futtermittel für landwirtschaftliche Nutztiere. DLG Verlag Frankfurt/a.M., 258–304.

Bogevik A.S., Samuelsen T.A., Aspevik T., Romarheim O.H., Aas T.S., Kalananthan T., Rønnestad I., 2021: Disintegration stability of extruded fish feed affects gastric functions in Atlantic salmon (*Salmo salar*). Aquaculture, 543, 737006.

Borquez A., Serrano E., Dantagnan P., Carrasco J., Hernandez A., 2011: Feeding high inclusion of whole grain white lupin (*Lupinus albus*) to rainbow trout (*Oncorhynchus mykiss*): effects on growth, nutrient digestibility, liver and intestine histology and muscle fatty acid composition. Aquaculture research, 42, 1067–1078.

Bowyer P.H., El-Haroun E.R., Salim H.S., Davies S.J., 2020: Benefits of a commercial solid-state fermentation (SSF) product on growth performance, feed efficiency and gut morphology of juvenile Nile tilapia (*Oreochromis niloticus*) fed different UK lupin meal cultivars. Aquaculture, 523. https://doi.org/10.1016/j.aquaculture.2020.735192.

Buraczewska L., Pastuszewska B., Smulikowska S., Wasilewko J., 1993: Response of pigs, rats and chickens to dietary level of alkaloids of different lupin species. Editors: van der Poel A.F.B., Huisman J., Saini H.S., 1993: Recent advances of research in antinutritional factors in legume seeds. Centre for agricultural publishing and documentation (PUDOC), Wageningen, 371–376.

Burel C., Boujard T., Tulli F., Kaushik S.J., 2000: Digestibility of extruded peas, extruded lupin, and rapeseed meal in rainbow trout (*Oncorhynchus mykiss*) and turbot (*Psetta maxima*). Aquaculture, 188, 285–298.

Carter C.G., Hauler R.C., 2000: Fish meal replacement by plant meals in extruded feeds for Atlantic salmon, *Salmo salar* L. Aquaculture, 185, 299–311.

Cheung P.C.-K., 1990: The carbohydrates of *Lupinus angustifolius*: a composite study of seeds and structural elucidation of the kernel cell-wall polysaccharides of *Lupinus angustifolius*, UNSW Sydney.

Chien Y., Chiu Y., 2003: Replacement of soybean (*Glycine max* (L.) Merrill) meal by lupin (*Lupinus angustifolius*) seed meal in diet for juvenile tilapia (*Oreochromis niloticus* × *O. aureus*) reared indoors. Aquaculture research, 34, 1261–1268.

Coenen M., Vervuert I., 2020: Getreidenachprodukte. In: Pferdefütterung. Georg Thieme Verlag Stuttgart, New York, 6th edition, 174–177.

David L.S., Nalle C.L., Abdollahi M.R., Ravindran V., 2024: Feeding value of lupins, field peas, faba beans and chickpeas for poultry: an overview. Animals, 14, 619–659. doi: 10.3390/ani14040619.

Durst L., Freitag M., Bellof G., 2021: Futtermittel für landwirtschaftliche Nutztiere. Anhang Tabelle 3, Aminosäuren: Körnerleguminosen. DLG Verlag FaM, 718–721.

Engel A.M., Klevenhusen F., Moenning J.-L., Numata J., Fischer-Tenhagen C., Sachse B., Schafer B., Fry H., Kappenstein O., Pieper R., 2022: Investigations on the transfer of quinolizidine alkaloids from *Lupinus angustifolius* into the milk of dairy cows. J. Agric. Food chem., 70, 11749–11758. https://doi.org/10.1021/acs.jafc.2c02517.

Erbas M., Certel M., Uslu M.K., 2005: Some chemical properties of white lupin seeds (*Lupinus albus* L.). Food chem., 89, 341–345.

Farhangi M., Carter C.G., 2001: Growth, physiological and immunological responses of rainbow trout (*Oncorhynchus mykiss*) to different dietary inclusion levels of dehulled lupin (*Lupinus angustifolius*). Aquaculture research, 32, 329–340.

Farrell D.J., Perez-Maldonado R.A., Mannion P.F., 1999: Optimum inclusion of field peas, faba beans, chick peas and sweet lupins in poultry diets. II. Broiler experiments. Br. Poult. Sci., 40, 674–680.

Gatlin D.M., Barrows F.T., Brown P., Dabrowski K., Gaylord T.G., Hardy R.W., Herman E., Hu G.S., Krogdahl Å., Nelson R., Overturf K., Rust M., Sealey W., Skonberg D., Souza E.J., Stone D., Wilson R., Wurtele E., 2007: Expanding the utilization of sustainable plant products in aquafeeds: a review. Aquaculture research, 38, 551–579.

Gdala J., 1998: Composition, properties, and nutritive value of dietary fibre of legume seeds. A review. J. Anim. Feed sci., 7, 131–149.

Gdala J., Jansman A.J.M., Van Leeuwen P., Huisman J., Verstegen M.W.A., 1996: Lupins (*L. luteus, L. albus, L. angustifolius*) as a protein source for young pigs. Anim. Feed Sci. Technol., 62, 239–249.

Glencross B., Boujard T., Kaushik S.J., 2003b: Influence of oligosaccharides on the digestibility of lupin meals when fed to rainbow trout, *Oncorhynchus mykiss*. Aquaculture, 219, 703–713.

Glencross B., Curnow J., Hawkins W., Kissil G., Peterson D., 2003a: Evaluation of the feed value of a transgenic strain of the narrow-leaf lupin (*Lupinus angustifolius*) in the diet of the marine fish, *Pagrus auratus*. Aquaculture nutrition, 9, 197–206.

Glencross B., Evans D., Dods K., Mccafferty P., Hawkins W., Maas R., Sipsas S., 2005: Evaluation of the digestible value of lupin and soybean protein concentrates and isolates when fed to rainbow trout, *Oncorhynchus mykiss*, using either stripping or settlement faecal collection methods. Aquaculture, 245, 211–220.

Glencross B., Hawkins W., Evans D., Rutherford N., Mccafferty P., Dods K., Karopoulos M., Veitch C., Sipsas S., Buichell B., 2008: Variability in the composition of lupin (*Lupinus angustifolius*) meals influences their digestible nutrient and energy value when fed to rainbow trout (*Oncorhynchus mykiss*). Aquaculture, 277, 220–230.

Glencross B., Hawkins W., Maas R., Karopoulos M., Hauler R., 2010: Evaluation of the influence of different species and cultivars of lupin kernel meal on the extrusion process, pellet properties and viscosity parameters of salmonid feeds. Aquaculture nutrition, 16, 13–24.

Glencross B.D., Rutherford N., Bourne N., 2012a: The influence of various starch and non-starch polysaccharides on the digestibility of diets fed to rainbow trout (*Oncorhynchus mykiss*). Aquaculture, 356–357, 141–146.

Glencross B.D., Booth M., Allan G.L., 2007: A feed is only as good as its ingredients – a review of ingredient evaluation strategies for aquaculture feeds. Aquaculture nutrition, 13, 17–34.

Glencross B.D., Carter C.G., Duijster N., Evans D.R., Dods K., Mccafferty P., Hawkins W.E., Maas R., Sipsas S., 2004: A comparison of the digestibility of a range of lupin and soybean protein products when fed to either Atlantic salmon (*Salmo salar*) or rainbow trout (*Oncorhynchus mykiss*). Aquaculture, 237, 333–346.

Glencross B.D., Rutherford N., Jones B., 2012b: Evaluation of the dietary inclusion of lupin (*Lupinus angustifolius*) kernel meal and lupin non-starch polysaccharide extract on the growth, feed utilization and nutrient digestibility of rainbow trout (*Oncorhynchus mykiss*). Aquaculture research, 43, 1057–1067.

Hoffmann M., Steinhöfel O., 2023: Futtermittel- und rationsspezifische Restriktionen. Agrimedia, Erling Verlag GmbH & Co. KG.

Hughes R.J., Kocher A., Choct M., 1998: Nutritive value of lupins for broilers. In: Proceedings of the 10th annual Australian poultry science symposium, Sydney, Australia, 8–10 February.

Jezierny D., Mosenthin R., Bauer E., 2010: The use of grain legumes as a protein source in pig nutrition: a review. Animal feed science and technology, 157, 111–128.

Kluth H., Boghun H., Engelhard T., Bulang M., Rodehutscord M., 2007: Bewertung von thermisch behandelten Lupinen als Rationskomponente für Hochleistungskühe. In: Heimische Körnerleguminosen mit geschütztem Protein in der Milchviehfütterung. UFOP-Schriften, 33, 63–87.

Kohajdova Z., Karovicova J., Schmidt S., 2011: Lupin composition and possible use in bakery – a review. Czech J. Food sci., 29, 203–211.

Lestingi A., Facciolongo A.M., Caputi Jambrenghi A., Ragni M., Toteda F., 2016: The use of peas and sweet lupin seeds alone or in association for fattening lambs: effects on performance, blood parameters and meat quality. Small ruminant research, 143, 15–23.

LfL Bayern, 2017: Gruber Tabelle zur Fütterung der Milchkühe, Zuchtrinder, Schafe, Ziegen. Bayerische Landesanstalt für Landwirtschaft, 42nd edition.

Martínez-Villaluenga C., Frías J., Vidal-Valverde C., 2006: Functional lupin seeds (*Lupinus albus L.* and *Lupinus luteus L.*) after extraction of α-galactosides. Food chem., 98, 291–299.

Mieczkowska A., Wroblewska M., Szymona J., 2005: Effect of dehulling and germination on chemical composition and protein characteristics of narrow-leafed lupin seeds. Food/Nahrung, 49, 91–97.

Molina-Poveda C., Lucas M., Jover M., 2013: Evaluation of the potential of Andean lupin meal (*Lupinus mutabilis* Sweet) as an alternative to fish meal in juvenile *Litopenaeus vannamei* diets. Aquaculture, 410–411, 148–156.

Nalle C.L., Ravindran V., Ravindran G., 2012: Nutritional value of white lupins (*Lupinus albus*) for broilers: apparent metabolisable energy, apparent ileal amino acid digestibility and production performance. Animal, 6, 579–585.

Olkowski B.I., Classen H.L., Wojnarowicz C., Olkowski A.A., 2005: Feeding high levels of lupine seeds to broiler chickens: plasma micronutrient status in the context of digesta viscosity and morphometric and ultrastructural changes in the gastrointestinal tract. Poult. Sci., 84, 1707–1715.

Olver M.D., Jonker A., 1997: Effect of sweet, bitter and soaked micronised bitter lupins on broiler performance. Br. Poult. Sci., 38, 203–208.

Pastuszewska B., 2016: Geschichte des Leguminosenanbaus. Editors: Jeroch H., Lipiec A., Abel H., Zentek J., Grela E.R., Bellof G., 2016: Körnerleguminosen als Futter- und Nahrungsmittel. DLG publishers, FaM, 10–22.

Pereira A., Ramos F., Sanches Silva A., 2022: Lupin (*Lupinus albus L.*) seeds: balancing the good and the bad and addressing future challenges. Molecules, 27, 8557.

Pereira T.G., Oliva-Teles A., 2004: Evaluation of micronized lupin seed meal as an alternative protein source in diets for gilthead sea bream *Sparus aurata* L. juveniles. Aquaculture Research, 35, 828–835.

Perez-Escamilla R., Vohra P., 1987: Lupins (*Lupinus albus var. Ulta*) replace a part of soybean meal in diet for growing chickens. Recent adv. Anim. Nutr., 9, 169.

Perez-Maldonado R.A., Mannion P.F., Farrell D.J., 1999: Optimum inclusion of field peas, faba beans, chickpeas and sweet lupins in poultry diets. Chemical composition and layer experiments. Br. Poult. Sci., 40, 667–673.

Petterson D.S., Mackintosh J.B., 1994: The chemical composition and nutritive value of australian grain legumes. grains research and development corporation (Editors: Canadian Grain Commission), https://www.researchgate.net/publication/202214404_The_Chemical_Composition_and_Nutritive_Value_of_Canadian_Pulses.

Pieper R., Gabel M., Pieper B., Souffrant W.B., Gruis D., 2007: Druckhydrothermisch behandelte Lupinen und Rapsextraktionsschrot in Rationen für Hochleistungskühe – ein möglicher Ersatz für Sojaextraktionsschrot. In: Heimische Körnerleguminosen mit geschütztem Protein in der Milchviehfütterung. UFOP-Schriften, 33, 50–62.

Pries M., Freitag M., Schöneborn A., Spiekers H., 2007: Effekte einer hydrothermischen Behandlung von Lupinen auf die Eiweißversorgung der Milchkuh. In: Heimische Körnerleguminosen mit geschütztem Protein in der Milchviehfütterung. UFOP-Schriften, 33, 44–49.

Rajeev R., Bavitha M., 2015: Lupins – an alternative protein source for aquaculture diets. International journal of applied research, 1, 4–8.

Refstie S., Glencross B., Landsverk T., Sørensen M., Lilleeng E., Hawkins W., Krogdahl Å., 2006: Digestive function and intestinal integrity in Atlantic salmon (*Salmo salar*) fed kernel meals and protein concentrates made from yellow or narrow-leafed lupins. Aquaculture, 261, 1382–1395.

Refstie S., Storebakken T., Baeverfjord G., Roem A.J., 2000: Replacement of fish meal with various plant protein sources in Atlantic salmon (*Salmo salar* L.) diets. Aquaculture.

Robaina L., Izquierdo M., Moyano F., Socorro J., Vergara J., Montero D., Fernandez-Palacios H., 1995: Soybean and lupin seed meals as protein sources in diets for gilthead seabream (*Sparus aurata*): nutritional and histological implications. Aquaculture, 130, 219–233.

Rodríguez-Estrada U., González-Alfaro K., Shene C., 2020: Replacement of fish meal by solid state fermented lupin (*Lupinus albus*) meal with 299v: effect on growth and immune status of juvenile Atlantic salmon (*Salmo salar*). Annals of animal science, 20, 991–1009.

Šajdler P., Zeman L., 2005: Porovnání stravitelnosti organických živin v krmných dávkách s obsahem lupiny a hrachu u koní. Acta Universitatis Agriculturae Et Silviculturae Mendelianae Brunensis Sborník Mendelovy Zemědělské A Lesnické Univerzity V Brně, LIII, 83–88.

Sami A.S., Schuster M., Schwarz F.J., 2010: Performance, carcass characteristics and chemical composition of beef affected by lupine seed, rapeseed meal and soybean meal. J. Anim. Physiol. Anim. Nutrition, 94, 465–473.

Samuelsen T.A., Hillestad M., Jacobsen H.J., Hjertnes T.J., Sixten H.J., 2021: Physical feed quality and starch content causes a biological response in Atlantic salmon (*Salmo salar* L). Aquaculture Reports, 21.

Serrano E., Lefillanca J.K., Carrasco J., Davies S.J., Hernandez A.A.J., 2024: Evaluation of andean lupin (*Lupinus mutabilis*) seed meal as a dietary component on growth performance, feed utilization, nutrient digestibility, and liver histology of rainbow trout (*Oncorhynchus mykiss*) Juveniles. Aquaculture reports, 34.

Silva F.C.P., Nicoli J.R., Zambonino-Infante J.L., Kaushik S., J.G.F., 2011: Influence of the diet on microbial diversity of faecal and gastrointestinal contents in gilthead sea bream (*Sparus aurata*) and intestinal contents in gold fish (*Carassius auratus*). FEMS Microbiol Ecol, 78, 285–296.

Smith D.M., Allan G.L., Williams K.C., Barlow C., 2000: Fishmeal replacement research for shrimp feed in Australia. Avances en Nutrición Acuícola V. Memorias del V Simposium Internacional de Nutrición Acuícola., Mérida, Yucatán, México.

Smith D.M., Tabrett S.J., Glencross B.D., Irvin S.J., Barclay M.C., 2007: Digestibility of lupin kernel meals in feeds for the black tiger shrimp, *Penaeus monodon*. Aquaculture, 264, 353–362.

Smith D.M., Tabrett S.J., Irvin S.J., Barclay M.C., 2006: Lupins – an alternative protein source for use in shrimp feeds. D.R.M.L. Elizabeth Cruz Suárez, Mireya Tapia Salazar, Martha G. Nieto López, David A. Villarreal Cavazos, Ana C. Puello Cruz y Armando García Ortega.

Stanford K., Mcallister T.A., Lees B.M., Xu Z.J., Cheng K.-J, 1996: Comparison of sweet white lupin seed, canola meal and soybean meal as protein supplements for lambs. Can. J. Anim. Sci., 76, 215–221.

Sudaryono, Tsvetnenko, Evans, 1999: Evaluation of potential of lupin meal as an alternative to fish meal in juvenile *Penaeus monodon* diets. Aquaculture nutrition, 5, 277–285.

Szczepański A., Adamek-Urbańska D., Kasprzak R., Szudrowicz H., Śliwiński J., Kamaszewski M., 2022: Lupin: a promising alternative protein source for aquaculture feeds? Aquaculture reports, 26, 101281.

Van Barneveld R.J., 1999: Understanding the nutritional chemistry of lupins (*Lupinus* spp.) seed to improve livestock production efficiency. Nutr. Res. Rev., 12, 203–230.

Van de Noort M., 2024: Chapter 11 – lupin: an important protein and nutrient source. Editors : S. Nadathur, J.P.D. Wanasundara, Scanlin L., 2024: Sustainable protein sources. Second Edition. Academic Press.

Van Vo B., Bui D.P., Nguyen H.Q., Fotedar R., 2015: Optimized fermented lupin (*Lupinus angustifolius*) inclusion in juvenile barramundi (*Lates calcarifer*) diets. Aquaculture, 444, 62–69.

Visvanathan R., Madhujith T., Gamage A., Zhang N., 2020: Lupin. Editors: Manickavasagan A., Thirunathan P., 2020: Pulses. Processing and product development. Springer Nature Switzerland AG, ISBN 978-3-030-41375-0, 169-203.

Watkins B.A., Mirosh L.W., 1987: White lupin as a protein source for layers. Poult. Sci., 66, 1798–1806.

Weber M., Krieg J., 2024: Einsatz unterschiedlicher Mengen an weißen Lupinen in der Schweinemast. Proteinmarkt, 05.06.2024.

Weiss M., Rebelein A., Slater M.J., 2020: Lupin kernel meal as fishmeal replacement in formulated feeds for the whiteleg shrimp (*Litopenaeus vannamei*). Aquaculture Nutrition, 26, 752–762.

Zeytin S., Hoerterer C., Weiss M., Slater M.J., 2018: Lupine meal as primary protein source in diets for European sea bass *Dicentrarchus labrax*. In preparation.

2.7 Soybeans (*Glycine max L.*) and products (European cultivation)

Gerhard Bellof

As already explained in section 2.1, soybeans (*Glycine max L.*) are increasingly being cultivated in Germany, particularly in southern Germany. In addition, more soybeans ("Donau Soja") are coming to Germany from south-eastern Europe. These soybeans fulfil the "non-GM cultivation" criterion. Reports show that European-grown soybeans offer a more sustainable option with significantly lower carbon emissions compared to imported soybeans.

Food retailers in Germany are increasingly demanding food from non-GM production. Some companies that process soybeans from European cultivation now also produce soya extraction meal in addition to soya cake. The Archer Daniel Midland Company (ADM), for example, also offers soybean meal with an increased crude protein content (HP-SES) from such non-GM soybeans – after dehulling – at its processing sites in Straubing (Bavaria) and Mainz (Rhineland-Palatinate).

In the following section, relevant feedstuffs from domestic (European) soybean cultivation will be presented. The focus here is on full-fat soybeans and the soybean cake produced from them.

Due to varietal differences, differing climatic conditions and cultivation methods, the value-determining ingredients of soybeans from European cultivation may differ from those imported from overseas. For this reason, the value-determining ingredients, feed value and recommendations for use in pig and poultry feed as well as cattle, sheep and goat feed will be compared below for full-fat soybeans from European cultivation and the soybean cakes and soybean meals produced from them.

2.7.1 *Ingredients*

2.7.1.1 Nutrients

Table 2.7.1-1 lists the contents of value-determining ingredients of soybeans (European). At 340 g crude protein/kg, soybeans have the highest protein content of all grain legumes after yellow sweet lupins (419 g CP/kg, 88% DM). However, the European soybeans analysed as part of the UFOP monitoring (Weber, 2018) show a wide range of variation in crude protein content (250–450 g/kg). The analysis of soybeans from European cultivation is therefore of increased importance.

Soybeans have by far the highest crude fat content of all grain legumes. The high fat content can limit the use of soybeans in animal feed. Compared to other oils and fats used in animal feed, soybean oil (fat) contains a significantly higher proportion of polyunsaturated fatty acids (table 2.7.1-2). As the fat composition of the feed is directly reflected in the fat quality of the body fat in pigs and poultry, the use of full-fat soybeans harbours the risk of excessively soft and unstable fat content in the carcass. This impairs the processability and shelf life of the products. For this reason, the product "soybean cake" made from soybeans with a maximum residual fat content of 10% is more suitable for use in animal feed. The fat removal caused by pressing leads to an enrichment of the other ingredients – including the proteins – in the cake.

The contents of the easily soluble carbohydrates starch and sugar are rather insignificant. As described for sweet lupins, the reported starch content should be regarded as an analytical artefact. The remarkably high NSP content (257 g/kg) does not lead to any adverse effects in young animal feeding.

2.7.1.2 Minerals

The soybeans have rather low calcium and sodium contents, but relatively high phosphorus values. In the by-products of soybeans, the calcium content increases slightly, while the phosphorus content decreases slightly. It should be noted that phosphorus is predominantly bound to phytin.

TABLE 2.7.1-1 Ingredients of soybeans (European cultivation)

Ingredient	Content (g/kg)		Min.–max
DM	880	1000	1000
CP	340	386	284–511
CL	200	227	159–273
CF	56	64	34–92
aNDFom	130	148	114–170
ADFom	90	102	80–114
STA	52	59	23–80

TABLE 2.7.1-1 Ingredients of soybeans (European cultivation) (*cont.*)

Ingredient	Content (g/kg)		Min.–max
SUG	71	81	68–102
NFE	237	269	–
ASH	47	53	51–60

WEBER (2018); LFL BAYERN (2014, 2017); JEROCH *ET AL.* (2016); DLG (2014)

FIGURE 2.7.1-2 Soybeans
PHOTO © UFOP/D. SCHNEIDER

TABLE 2.7.1-2 Fatty acid pattern of vegetable oils (mean proportions in %)

Fatty acids	Soybean	Rapeseed	Sunflower	Linseed	Palm	Coconut
Saturated fatty acids	6.5	7	8	5.9	1.5	83
Sum C8–C14	4.5	–	2	2.4	45	7
Palmitic acid (C16)					4.7	5
Stearic acid (C18)						

TABLE 2.7.1-2 Fatty acid pattern of vegetable oils (mean proportions in %) (*cont.*)

Fatty acids	Soybean	Rape-seed	Sunflower	Lin-seed	Palm	Coconut
Unsaturated fatty acids	27	61	16	21	39	5
Oleic acid (18:1)	50	21	71	18	9	1
Linoleic acid (18:2)	5	11	1	50	0.3	0.2
α-linolenic acid (18:3; ω-3)						
Melting point, °C	−12	−13	−17	−20	25–35	23

DURST *ET AL.* (2021)

2.7.1.3 Antinutritive factors (ANF)

ANF – such as: Tannins (tannins), protease inhibitors (inhibitors), lectins and saponins – can be found in grain legumes (section 2.3).

Trypsin inhibitors are particularly important in soybeans and their by-products.

These substances can inhibit the action of the protein-splitting enzyme trypsin in the small intestine. Before feeding soybeans and their processed products to poultry or pigs, it is therefore necessary to thermally inactivate the protease inhibitors they contain. In practice, this is carried out either "decentralised" in stationary or mobile toasting plants. Centralised soybean processing in oil mills for vegetable oil production is usually linked to a toasting process step that ensures defined toasting conditions.

However, thermal treatment also harbours the risk of protein damage. A compromise must therefore be sought between the positive effects (elimination of performance-inhibiting ingredients and enzymes that impair shelf life, gentle denaturation of the protein bodies) and the onset of protein-damaging reactions. Even a slight excess of temperature can lead to damage and a reduction in the content of the sulphur-containing amino acids cysteine and methionine, as well as the amino acid lysine.

A number of simple analytical methods have been developed to check the proper pretreatment of soybeans, such as the determination of urease activity, cresol red absorption and protein solubility. The direct determination of trypsin inhibitor activity (TIA) can also be carried out according to the official A.O.C.S. method (1990). The activity of the inhibitor is expressed in mg trypsin inhibitor per g dry matter (mg TIA/g DM). Thurner (2020) recommends maintaining a target value of less than 3 mg TIA/g DM soybean cake. In complete feed mixtures for broilers, a value of less than 2 mg TIA/g DM should be aimed for. The determination of urease activity is used to indirectly record the inhibitor effect, as direct measurement is comparatively time-consuming. Therefore, the residual activity of another ingredient characteristic of the soya bean, the urease enzyme, is measured as a substitute parameter. For optimally toasted soya products, a urease activity between 0.4 mg N/g/min. and the detection limit is required. After reaching 100 °C, the urease activity drops very quickly to low values, the changes of which are meaningless. This means that only unheated batches can be identified.

The protein solubility in water (PDI) is another common criterion for testing the effect of heat treatment. According to Naumann and Bassler (1988), an optimum range of 10 to 35% can be assumed for soya products. Puntigam *et al.* (2021) state a target range of 10–25%. In addition to the protein solubility of the protein in water (PDI), the protein solubility in potassium hydroxide solution (KOH) is often determined as a further parameter. Recent studies by Thurner (2020) show that the lower limit – heat damage to the proteins – can be set at 70% for KOH. Undertreatment occurs at KOH values of > 85%.

TIA and PDI or KOH should be determined for a clear assessment of the heat treatment. Such characteristic values are shown in table 2.7.1-3. Compared to untreated soybeans, the heat-treated soya products shown there are within the respective target range in terms of protein solubility (PDI, KOH). However, the TIA values for soybeans and soybean cakes are above the target range of < 3 g/kg DM.

TABLE 2.7.1-3 Relevant ingredients and characteristic values for the heat treatment of soya products from European cultivation compared to soybean meal (overseas) (in g/kg or %)

Ingredient/ Characteristic value		**Soybeans (Crude)**[1]	**Soybeans (Thermal treatment)**[1]	**Soybean cake (Thermal treatment)**[1]	**Soybean meal (Europe) (Thermal treatment)**[1]	**Soybean mea[l] (Oversea) (Thermal treatment)**[2]
DM	g	880 /910	880 / 941	880 / 935	882	880
CP	g	359 / 371	371 / 397	410 / 436	436	442
CL	g	197 / 204	192 /206	107 /114	18	21
CF	g	55 / 57	50 / 53	47 / 50	55	49
Protein solubility (PDI)	%	59	24	18	14	(13)[1]
Protein solubility (KOH)	%	94	82	85	80	(76)[1]
Trypsin inhibitor activity (TIA)	g/kg DM	26.2	6.9	7.9	1.1	(2.4)[1]

1 Puntigam *et al.* (2021).
2 Durst *et al.* (2021).

In practice, there are reports that the soya products processed in decentralised plants do not always meet the requirements described. There is a need for optimisation so that the standard set by soybean meal can also be consistently achieved for domestic soya products.

2.7.2 *Feed value*

2.7.2.1 Energy

Soybeans also have a very high energy content for pigs and poultry due to their high protein and fat content. These are significantly higher than those of wheat or grain maize. The same applies to poultry feed.

Soybeans are also an extremely energy-rich feed for ruminants. However, the high fat content has a limiting effect here due to the limited fat tolerance.

2.7.2.2 Protein

The heat treatment of soybeans leads to a significant increase in the UDP content. In German feed value tables (LfL Bayern, 2017), a UDP content of 20% is specified. According to Sauvant *et al.* (2004), the UDP content for the protein of toasted soybeans is approx. 37%, for extruded soybeans even 53% and thus at least at the level of conventional soybean meal. In principle, soybeans can also be used for ruminants without heat treatment. However, the UDP content is then a maximum of

TABLE 2.7.2-1 Amino acid content and digestibility (pigs, poultry) of soybeans (European cultivation)

Amino acid	Content			Digestibility (pcd) in %	
	g/kg (88% DM, 340 g CP)	g je 100 g CP	g je 100 g CP (min.–max.)	Pig	Poultry
Lys	21.8	6.4	5.0–14.5	87	88
Met	4.8	1.4	1.4–9.1	85	85
Cys	5.0	1.5	1.4–10.8	80	77
Thr	13.4	3.9	3.8–19.7	84	85
Trp	4.8	1.4	0.8–2.5	77	n.d.
Ile	16.2	4.8	n.d.	85	87
Leu	26.2	7.7	n.d.	85	87
Val	16.8	4.9	n.d.	84	86

WEBER (2018); LFL BAYERN (2014, 2017) JEROCH *ET AL.* (2016); DLG (2014); MOSENTHIN ET. AL. (2004); SAUVANT *ET AL.* (2004); BELLOF *ET AL.* (2020)

10%. This means that considerable potential is wasted. This applies in particular to high-yielding ruminants with an increased need for rumen-stable protein.

The contents of the most important amino acids in soybeans are shown in table 2.7.2-1. Measured against the requirements of poultry and growing pigs, soya protein is too low in methionine. In contrast, the subordinate essential amino acids in particular are present in sufficient quantities. Soybeans can therefore ensure the supply of these amino acids in pig and poultry rations in a similar way to soya extraction meal.

The amino acid digestibilities for pigs are at a high level. Only the amino acid tryptophan has a digestibility of less than 80%. High amino acid digestibilities were also found for poultry. With the exception of cysteine, all values are at least 85% (table 2.7.2-1).

2.7.3 *Feeding recommendations for soybeans and soybean cake*

Soybeans from European cultivation are used in particular in organic feeding. Prior heat treatment is mandatory for use in monogastric feeding and recommended for ruminant feeding (2.7.2). Proper heat treatment is therefore an important prerequisite for successful feeding. The high fat content usually limits its use. The limits of use can be shifted by partial fat removal. Soya cake made from domestic soybeans is therefore particularly important for organic feeding. The following explanations therefore also contain recommendations for the use of soya cake (with a residual fat content of 10% and a crude protein content of 395 g/kg, at 88% DM).

2.7.3.1 Cattle and small ruminants

In a feeding trial with dairy cows, Ettle *et al.* (2011) tested the use of differently pre-treated soybeans (dried versus toasted). The cows were given ad libitum partial mixed rations. These contained dried or toasted full-fat soybeans (8.1% in the DM). Above milk yields of 24 kg/cow and day, an additional milk yield feed was given. The cows consumed an additional 0.9 kg of soya extraction meal per cow per day via this mixture. With the partial mixed rations, which contained 4.4–4.8% CL (in the DM), the cows consumed around 1.5 kg of full-fat beans. There were no different effects of dried or toasted soybeans on the milk yield of the cows. However, the average milk yield in this trial was only 28.5 kg per cow per day.

Mogensen *et al.* (2008) also compared the treatment effect of soybeans in a dairy cow feeding trial under organic farming conditions. The fresh lactating cows were fed either 3.4 kg of raw soybeans per day or 3.5 kg of toasted soybeans as the sole protein feed. There were clear differences in milk yield between the two groups. For example, the cows fed toasted soybeans achieved a significantly higher milk yield (30.2 kg versus 29.1 kg/cow and day) and tended to have higher milk constituents. However, the low milk fat contents (3.80% versus 3.66%) in both groups indicate that the high feed fat intake led to milk fat depression.

TABLE 2.7.3-1 Recommendations for the use of soybeans and soybean cake in cattle and sheep feeding (in kg/animal/day or mixture proportion in % of the concentrate mixture)

Animal group	Soybeans	Soybean cake
Cattle		
Dairy cows (lactate)	Max. 2.5 kg	Max. 4 kg
Beef cattle	1.5 kg	2 kg
Sheep		
Ewe (lactate)	15%	20%
Lambs	15%	20%

BELLOF *ET AL.* (2013); COMPLETED

In a feeding trial, Facciolongo *et al.* (2014) tested the suitability of typical grain legumes as an essential source of protein for growing ram lambs (weight range 12 to 21 kg live weight). Four different dry feed mixtures (concentrate and coarse feed mixed and pelleted, for free intake) were prepared containing either field beans (25%), peas (25%), lupins (25%) or soybeans (12%). The rations were isoenergetic and – with the exception of the pea mixture – isonitrogenous. The protein was balanced with lucerne meal. The lambs fed with field beans achieved significantly higher daily weight gains than the animals in the soya bean group (210 g versus 160 g). The daily weight gains in the pea and lupin groups were in between. No directional differences in carcass value were found between the feeding groups. The soybeans used in the trial were not subjected to any heat treatment. The low daily weight gains of the lambs fed with these soybeans emphasise the need to ensure sufficiently high UDP levels, especially in young lambs.

The recommendations for the use of soybeans and soya cake in cattle and sheep feed are summarised in table 2.7.3-1.

There are only a few studies on the use of soybeans for goat feeding in Central Europe. The recommendations given in table 2.7.3-1 for sheep feeding can be used for dairy goats (lactating) and kid rearing (fattening).

2.7.3.2 Pigs

Pigs generally have a good fat tolerance. Feed with a high fat content can therefore be used. In pigs, there is a close correlation between the fat composition of the feed and the fat quality of the body fat. The use of full-fat soybeans therefore harbours the risk of insufficient fat quality in the carcass. This can lead to problems during processing and limit the shelf life of products made from them. Soybeans should therefore only be used to a limited extent in pig fattening. On the other hand, higher proportions of soybean cake can be recommended in complete feed mixtures (table 2.7.3-2).

TABLE 2.7.3-2 Recommendations for the use of soybeans and soybean cake in pig feed (maximum proportions for complete feed mixtures, figures in %)

Animal group	Soybeans	Soybean cake
Piglets		
≤ 20 kg	10	20
> 20 kg	15	20
Fattening pigs		
Starter	15	15
Finisher	10	15
Sows		
Pregnant	5	6
Lactate	20	20

WEBER *ET AL.* (2020)

2.7.3.3 Poultry

Due to the lower fat content and the resulting lower polyenoic acid intake as well as the higher protein content, soya cake is the preferred protein feed for poultry fattening (Steiner and Bellof, 2009; Bellof, 2013). This is reflected in the higher use recommendations for both broiler and turkey fattening (table 2.75.3-3).

TABLE 2.7.3-3 Recommendations for the use of soybeans and soybean cake in poultry feed (maximum mixture proportions for complete feed mixtures, figures in %)

Animal group	Soybeans	Soybean cake
Laying hens		
Egg production	15	20
Reproduction	15	20
Broiler		
Starter (≤. 4. LW)[1]	15	20
Finisher (> 4. LW)	15	20
Fattening turkeys		
Rearing (P[2] 1 / P2)	10	20
Fattening (P 3–4)	10	20
Fattening (P 5–7)	10	15

1 LW: week of life.
2 P: phase.

BELLOF *ET AL.* (2020); COMPLETED

2.7.3.4 Fishes and shrimps

Sinem Zeytin, Vanessa Fuchs and Monika Weiß

Soybean meal (SBM) protein content in Europe generally varies but is typically around 42% crude protein to over 65–70% protein in soy protein concentrate (SPC), making it an attractive protein source for fish and shrimp diets. Since the beginning of 1950 and 1960 as the aquaculture industry expanded and the demand for sustainable and cost-effective protein sources increased, SBM is included in aquafeeds where its protein content and amino acid profile complement other ingredients to meet the nutritional requirements of the target species (Hardy, 2000; Dersjant-Li, 2021). From 1990s onwards, the use of SBM became a standard ingredient in many commercial aquafeed formulations as fishmeal comes from a highly limited and unsustainable source.

In modern aquaculture, advances in processing and formulation techniques improved soybeans efficacy and acceptance in aquafeeds. Despite all efforts, SBM is currently considered less sustainable in aquafeeds and potential alternatives have been explored. Each alternative source such as insect meal, algal meal, single-cell proteins and other plant-based proteins comes with its own set of benefits and challenges, and ongoing research and development are continuing to make these alternatives viable at scale. The scale and supply of soybeans are stable however, like other plant-derived protein sources, soybeans have anti nutritive factors (ANFs) that can reduce palatability, protein utilization, or growth (Hardy and Kaushik, 2021). Soybean may replace animal protein in diets for aquatic animals to a certain extent. However, with increasing substitution of e.g. fishmeal by SBM the performance of fish decline. SBM has been found to affect digestive functions of several species, especially in salmonids (Krogdahl *et al.*, 2010).

Salmon and trout have higher protein requirements and tolerate less plant proteins and ANFs, requiring careful formulation and sometimes supplementation with amino acids (e.g., methionine), enzyme supplementation (e.g., phytase) and other nutrients. Hence, the inclusion levels of SBM in their diets are generally lower. On the other hand, tilapia and carp can utilize plant proteins more effectively, allowing for higher inclusion rates of SBM.

SBM is usually referred to as fat-extracted, dehulled soybeans ground to a meal. SPC is the most extensively used soy product in feeds for salmonids as it has lower content of ANFs, soluble carbohydrates and fibre and therefore can be included at higher levels. In 2010, the inclusion level in commercial feeds ranged between 17–22% for salmonids in Norway. However, in Atlantic salmon, some ANFs in soybeans induce intestinal enteritis and changes in the composition of the gut microbiota, which affects gut health and immunity as well as feed digestibility and fish growth. Phytic acid can be reduced by pretreating the SPC with phytase.

Historic development of the application of soy products began with full fat soybeans, progressed by dehulling and strongly improved for most species and a more flexible inclusion of products with defatted SBM. The production of concentrates allowed even higher inclusion rates, especially for species of higher trophic levels, requiring higher protein inputs. Latest studies concentrate on fermentation of SBM with *Bacillus* (McLean *et al.*, 2020, Dan *et al.*, 2022, Abd El-Naby *et al.*, 2024) species, which further improves opportunities of soy inclusion.

Today aquaculture relies on a mixture of plant protein ingredients such as corn gluten, wheat gluten, sunflower meal, chickpeas, lentils, lupine, faba beans and many other legumes that can replace soy and fish meal in many formulations that are more efficiently utilized by the fish (De Santis *et al.*, 2016, Bonvini *et al.*, 2018). The overall composition of the diet, including the balance of amino acids, lipids, and carbohydrates, affects inclusion rates of ingredients.

Additionally, newer studies attempt to overcome negative effects soybean inclusion has on fishes, using functional ingredients and pre-treatments like heat, soaking, enzyme treatment, fermentation or heat processing e.g. the extrusion process. Adding essential amino acids, vitamins, and minerals can help achieve balanced nutrition when SBM is included at higher rates. In recent studies, attention is also paid to appetite-suppressing substances (Dan *et al.*, 2022). An overview of new studies with recommended inclusion levels of soybeans in feeds for different fish and shrimp species is given in table 2.7.3-4.

In rainbow trout (*Oncorhynchus mykiss*) diets, fermented or phytase-treated SBM can be applied at levels of 30 to 48% without compromising fish performances or physiological abnormalities (Yamamoto *et al.*, 2010, Yang *et al.*, 2011, Barnes *et al.*, 2014). In comparison, the inclusion of SPC is possible at levels of 32 to even 62% (100% replacement of fishmeal (FM)) in rainbow trout diets (Kaushik *et al.*, 1995, Stickney *et al.*, 1996, Mambrini *et al.*, 1999, Vielma *et al.*, 2000).

One study showed that a diet with a well balanced mix of plant-proteins, including 23.5% SPC per kg diet, and highly available inorganic phosphate can replace 100% FM with no apparent reduction in feed utilization and growth of rainbow trout (Lee *et al.*, 2010). In general, the recommended inclusion level of soy protein highly depends on the overall nutritional composition of the feed and, for instance, the addition of essential amino acids, phosphorus or phytase to compensate for nutrient deficiencies.

Diets of Atlantic salmon (*Salmo salar*) parr can contain ~27% SBM and ~38% to 45% SPC without reducing growth performance, nutrient utilization or health (Carter and Hauler, 2000; Metochis *et al.*, 2016a; Metochis *et al.*, 2016b). In diets for post-smolt Atlantic salmon, SBM can be included at low levels of ~11 to 13% without negative effects on feed conversion or fish growth (Krogdahl *et al.*, 2003; Torstensen *et al.*, 2008). Even low SBM inclusion levels already caused moderate pathohistological/morphological changes in the intestine of salmons. Higher inclusion rates > 20% showed to reduce fish performances and induce severe

TABLE 2.7.3-4 Studies with recommended inclusion levels of soybeans in feeds for different fish and shrimp species

Species	Initial species weight (g)	Product	Replaced resource	Inclusion rates (%)	Recommendable inclusion rate (%)	Reference
Atlantic salmon (*Salmo salar*)	~47	SBM	FM	20.4 and 27.3	27.3	Carter and Hauler, 2000
parr	~9	SPC	FM	28.8, 45.4 and 59.9	45.4	Metochis *et al.*, 2016b
	~29	SPC	FM	38.4, 48.9 and 59.1	38.4	Metochis *et al.*, 2016a
Atlantic salmon post-smolt	~280	SBM	FM	7.6, 11.7, 15.3, 19.4 and 27.0	11.7	Krogdahl *et al.*, 2003
	~355 to > 3900	SBM	FM	13 (4 mm pellet size), 11 (6 mm pellet size)	13, 11	Torstensen *et al.*, 2008
	~187	Hulled SBM, WF = white flakes (dehulled SBM), fermented WF, BPSBM = biotechnologically processed SBM	FM	WF: 16.6 and 33.1; fermented WF: 14.6 and 28.8; BPSMB: 15.3 and 30.8; SBM: 17.5	Fermented WF: up to 28.8; BPSMB: up to 30.8	Refstie *et al.*, 2005
	~100	SPC, phytase-treated SPC	FM	48	Phytase-treated SPC: 48	Storebakken *et al.*, 1998
	~270–305	SBM, dehulled SBM, dehulled full-fat SBM, SPC	FM	Full-fat SBM: 60, SBM: 51.2, dehulled SBM: 46.8, SPC: 37.6	SPC: 37.6	Olli *et al.*, 1994

TABLE 2.7.3-4 Studies with recommended inclusion levels of soybeans in feeds for different fish and shrimp species (*cont.*)

Species	Initial species weight (g)	Product	Replaced resource	Inclusion rates (%)	Recommendable inclusion rate (%)	Reference
Rainbow trout (*Onkorhynchus mykiss*)	~4	Phytase-pretreated SBM	FM	13.6, 27.2, 40.8 and 54.4	Up to 40.8	Yang *et al.*, 2011
	~6.1	Fermented SBM	FM	10, 20, 30, 40 and 50	Up to 30	Barnes *et al.*, 2014
	~15	SPC	FM	15.9, 31.8, 47.7 and 63.7	31.8 (50% FM replacement)	Stickney *et al.*, 1996
	~106	SPC	FM	32.0, 49.0, 63.5, 63.4 (+ methionine) and 63.2 (+ methionine)	32.0	Mambrini *et al.*, 1999
	~250–2000	SPC and SBM	FM	31.5 SPC, 12.1 SBM; 69% of the protein from soy ingredients	31.5 SPC, 12.1 SBM	Vielma *et al.*, 2000
	~31.5	SPC, CG, WG, fish bone meal	FM	21.0–24.3	Plant proteins: 23.5 SPC, 21.0 CG, 4.5 WG (no FM)	Lee *et al.*, 2010

TABLE 2.7.3-4 Studies with recommended inclusion levels of soybeans in feeds for different fish and shrimp species (*cont.*)

Species	Initial species weight (g)	Product	Replaced resource	Inclusion rates (%)	Recommendable inclusion rate (%)	Reference
Turbot (*Scophthalmus maximus*)	~9	SBM	FM	26, 35 and 44	35	Peng *et al.*, 2013
	~95.8	SPC	FM	25		Fuchs *et al.*, 2015
	~7.3	Fermented SBM	FM	~30		Jiang *et al.*, 2019
	~5.6	Extruded full-fat soybean	FM	~7, ~14, ~21	~14	Zhang *et al.*, 2023
	~5.3	Fermented	FM	~32–34	34 (fermented SBM)	Dan *et al.*, 2022
Starry flounder (*Platichthys stellatus*)	~5.6	Soy protein hydrolysate	FM	14, 23.2, 37.7, 49.7, 61.7, 71.9	38.32* (optimum)	Song *et al.*, 2014
Olive flounder (*Paralichthys olivaceus*)	~6.2	Extruded SBM	FM	8, 16, 24, 32	32	Saitoh *et al.*, 2003
European sea bass (*Dicentrarchus labrax*)	110	SBM	FM	35, 50, 70	50	Altan *et al.*, 2010
	188	SBM (toasted, dehulled, subjected to dry extrusion, mechanical oil extraction and enzyme-treatment)	FM	25, 50, 60	50% of fish meal protein (i.e., 40% of the formula weight)	Tibaldi *et al.*, 2006
	69	SBM		10, 20, 30	30	Bonvini *et al.*, 2018

TABLE 2.7.3-4 Studies with recommended inclusion levels of soybeans in feeds for different fish and shrimp species (*cont.*)

Species	Initial species weight (g)	Product	Replaced resource	Inclusion rates (%)	Recommendable inclusion rate (%)	Reference
Gilthead seabream (*Sparus aurata*)	12	SPC	FM	20, 40, 72.5	Up to 40	Kissil *et al.*, 2000
	Experiment I: 6 Experiment II: 1.6	Exp I: SBM Exp II: full-fat meal	FM	Exp I: 10, 20, 30, 40 Exp II: 35	Exp I: up to 20 Exp II: properly heated full-fat SBM	Nengas *et al.*, 1996
Black sea bream (*Acanthopagrus schlegeli*)	1.2	Fermented SBM	FM	8, 16, 24, 32	24	Azarm *et al.*, 2014
Red sea bream (*Pagrus major*)	24	SBM	FM	30	30	Biswas *et al.*, 2007
	Exp I: 11.7 Exp II: 178	SPC		52	52 SPC	Takagi *et al.*, 2001
	7.3	SBM (dehulled)	FM	30, 40, 47, 56	56 (no FM)	Kader *et al.*, 2012
Asian sea bass (*Lates calcarifer*)	1.3	SBM (extruded full-fat, steamed and soaked raw full-fat)	FM	21, 27, 28.5, 27.5	21, 27, 28.5	Boonyaratpalin *et al.*, 1998

TABLE 2.7.3-4 Studies with recommended inclusion levels of soybeans in feeds for different fish and shrimp species (*cont.*)

Species	Initial species weight (g)	Product	Replaced resource	Inclusion rates (%)	Recommendable inclusion rate (%)	Reference
Japanese seabass (*Lateolabrax japonicus*)	7.1	Fermented SBM	FM	20, 40, 80	Up to 40	Rahimnejad *et al.*, 2019
Pacific white shrimp (*Litopenaeus vannamei*)	~0.03	SBM	FM	32.5, 34.8, 37.2, 39.5	39.5 (no FM)	Amaya *et al.*, 2007
	~10.2	SBM (different sources)	Reference diet	30	30, depending on source	Galkanda-Arachchige *et al.*, 2020
	~2.7	SBM	FM	34	< 34	Zhang *et al.*, 2020
	~0.23	SBM	FM	3.1, 6.2, 9.2, 12.3, 15.4, 18.5	15.4	Ray *et al.*, 2020
	~1.6	Fermented SBM	FM	20, 30, 40	20	Abd El-Naby *et al.*, 2024

TABLE 2.7.3-4 Studies with recommended inclusion levels of soybeans in feeds for different fish and shrimp species (*cont.*)

Species	Initial species weight (g)	Product	Replaced resource	Inclusion rates (%)	Recommendable inclusion rate (%)	Reference
Pacific white shrimp (*Litopenaeus vannamei*)	PL 1	SBM, fermented SBM	FM	26.1, 44.1	44.1 (no explicit growth data)	McLean *et al.*, 2020
	~0.7	SBM, fermented SBM	FM	20, 40	SBM: 20 Fermented SBM: 40	Bae *et al.*, 2020
	~0.2	SMB, fermented (F)SBM	FM	SBM: 17.7, 35.3, 52.9 FSBM: 14,9, 29.7, 44.4, 59.2, 74.0	SBM: 37.4 FSBM: 61.7	Shiu *et al.*, 2015

SBM = soybean meal; SPC = soy protein concentrate; FM = fishmeal; CG = corn gluten; WG = wheat gluten, * = calculated optimum inclusion morphological changes and enteritis (Baeverfjord and Krogdahl, 1996; Urán *et al.*, 2008). Fermented or biotechnologically processed SBM that is reduced in oligosaccharides and ANFs may be included at higher levels of 28 to 31% in post-smolt salmon diets (Refstie *et al.*, 1998; Refstie *et al.*, 2005). Similarly, SPC and particularly phytase-treated SPC is a good alternative to SBM and can be added with a rate of 38 to 50% without impairing performances in fish (Olli *et al.*, 1994; Storebakken *et al.*, 1998; Storebakken *et al.*, 2000). In practical applications, incorporating SPC into aquafeeds in Atlantic salmon has shown that diets with up to 30% SPC and the addition of 60% of insect meal can enhance feed efficiency while minimizing inflammation issues often associated with higher soy inclusion level (Belghit *et al.*, 2018; Li *et al.*, 2022).

Petereit *et al.* (2022) is a great example of a current goal of aquafeed feed formulation which is not focused not on a single ingredient analysis, but rather feed formulation concepts based on combinations of more sustainable ingredients to test for growth, nutritional, welfare and sensory parameters. The study examined whether the replacement of fishmeal, fish oil and soy products in feed efficiency by alternative ingredients affects the performance of European sea bass (*Dicentrarchus labrax*). It presents future feed formulations for farmed European sea bass that would replace the fishmeal completely with hydrolysates, exchange the SBM with more sustainable plants, such as wheat and enriched algae, and add agriculture byproducts, such as feather meal and others, for a better mineral composition and digestibility.

For formulation of practical diets in on-growing of European sea bass, SBM up to 30% can be successfully incorporated into feeds containing low fishmeal inclusion and combination of plant ingredients without any negative effect on health (Bonvini *et al.*, 2018).

In early studies with Gilthead sea bream (*Sparus aurata*), fishmeal replacement up to 20% SBM did not affect growth, feed or protein efficiency. However, heat-treated full-fat SBM appeared to be a good protein source for sea bream (Robaina *et al.*, 1995; Nengas *et al.*, 1996). Later studies showed that inclusion rates up to 40% of SPC did not affect growth performance with a decrease of fishmeal content in the formulated diets for sea bream however, it is recommended to identify the possible amino acid deficiencies in the SPC. As (Takagi *et al.*, 2001; Watson *et al.*, 2013; Kotzamanis *et al.*, 2020) reported that supplementation of lysine, methionine and taurine in high plant protein-based diets is beneficial to improve the growth performance of piscivorous fish species. In another study from Kissil *et al.* (2000) where two sizes of sea bream (small 9 g and large 50 g) found that extrusion of soy affected differently large and small fish improving both protein productive value and feed conversion ratio of large fish, but only protein productive value of small fish (Venou *et al.*, 2006).

Studies conducted with turbot (*Scophthalmus maximus*) revealed, that dietary supplementation with a variety of functional ingredients like pre- and probiotics, nucleotides, *Bacillus* strains or polysaccharides, which are all reported to have a positive effect on fish performance, failed to improve growth, when fed with a plant protein/ soy protein-based diet (Peng *et al.*, 2013; Fuchs *et al.*, 2015). Fermentation of the SBM is obviously the means of choice to improve digestibility and successful inclusion rates. For example with full fat soybeans an inclusion rate of 14% was successful in turbot juveniles (Zhang *et al.*, 2023), but the fermentation of SBM allows an inclusion rate of 34% resulting in better growth, an improved antioxidant capacity and less inflammations compared to non-fermented SBM (Dan *et al.*, 2022).

A vast variety of studies exist about the application of soy for shrimp, most of them focusing on the whiteleg shrimp (*Litopenaeus vannamei*). Generally, SBM can replace 37%–56% of the FM in the feed, but depending on the production system (Clearwater vs. Biofloc) even a FM replacement of 100% was achieved (Amaya *et al.*, 2007). Newer studies often focus on *Bacillus subtilis* fermented soy bean products, which have higher crude protein, amino acids and free amino acids compared to SBM, which allows very high inclusion rates of the fermented SBM up to 62% in the diet (Shiu *et al.*, 2015).

References

Abd El-Naby A.S., Eid A.E., Gaafar A.Y., Sharawy Z., Khattaby A.A., El-sharawy M.S., El Asely A.M., 2024: Overall evaluation of the replacement of fermented soybean to fish meal in juvenile white shrimp, *Litopenaeus vannamei* diet: growth, health status, and hepatopancreas histomorphology. Aquaculture international, 32, 1665–1683.

Altan O., Gamsiz K., Korkut A.Y., 2010: Soybean meal and rendered animal protein ingredients replace fishmeal in practical diets for sea bass. israeli journal of aquaculture – Bamidgeh, 62(1), 56–62. WOS:000277890800008.

Amaya E.A., Davis D.A., Rouse D.B., 2007: Replacement of fish meal in practical diets for the Pacific white shrimp (*Litopenaeus vannamei*) reared under pond conditions. Aquaculture, 262, 393–401.

Azarm H.M., Lee S.M., 2014: Effects of partial substitution of dietary fish meal by fermented soybean meal on growth performance, amino acid and biochemical parameters of juvenile black sea bream. Aquaculture research, 45(6), 994–1003. doi.org/10.1111/are.12040.

Bae J., Hamidoghli A., Djaballah M.S., Maamri S., Hamdi A., Souffi I., Farris N.W., Bai S.C., 2020: Effects of three different dietary plant protein sources as fishmeal replacers in juvenile whiteleg shrimp, Litopenaeus vannamei. Fisheries and Aquatic Sciences 23(1), 2.

Baeverfjord G., Krogdahl Å., 1996. Development and regression of soybean meal induced enteritis in Atlantic salmon, *Salmo salar* L., distal intestine: a comparison with the intestines of fasted fish. Journal of fish diseases 19(5): 375–387.

Barnes M.E., Brown M.L., Bruce T., Sindelar S., Neiger R., 2014: Rainbow trout rearing performance, intestinal morphology, and immune response after long-term feeding of high levels of fermented soybean meal. North American journal of aquaculture, 76, 333–345.

Belghit I., Liland N.S., Waagbo R., Biancarosa I., Pelusio N., Li Y.X., Krogdahl Å., Lock E.J., 2018: Potential of insect-based diets for Atlantic salmon (*Salmo salar*). Aquaculture, 491, 72–81.

Bellof G., 2013: Heimische Sojaprodukte in der Fütterung landwirtschaftlicher Nutztiere. Soja-Fütterungsfibel, Editor: FibL Deutschland. https://orgprints.org/id/eprint/24970/1/soja_fuetterungsfibel.pdf. Retrieved: August 7, 2024.

Bellof G., Aulrich K., Weiß J., 2013: Körnerleguminosen in der Fütterung. In: Körnerleguminosen anbauen und verwerten. KTBL-Heft 100, 36–41.

Bellof G., Halle I., Rodehutscord M., 2020: UFOP-Praxisinformation: Ackerbohnen, Körnerfuttererbsen, Süßlupinen und Sojabohnen in der Geflügelfütterung. Editor: Union zur Förderung von Öl- und Proteinpflanzen e.V. (UFOP), Berlin, 2. aktualis. Auflage.

Biswas A.K., Kaku H., Ji S.C., Seoka M., Takii K., 2007: Use of soybean meal and phytase for partial replacement of fish meal in the diet of red sea bream. Aquaculture, 267(1–4), 284–291. Doi:10.1016/j.aquaculture.2007.01.014.

Bonvini E., Bonaldo A., Mandrioli L., Sirri R., Dondi F., Bianco C., Fontanillas R., Mongile F., Gatta P.P., Parma L., 2018: Effects of feeding low fishmeal diets with increasing soybean meal levels on growth, gut histology and plasma biochemistry of sea bass. Animal, 12, 923–930.

Boonyaratpalin M., Suraneiranat P., Tunpibal T., 1998: Replacement of fish meal with various types of soybean products in diets for the Asian seabass, *Lates calcarifer*. Aquaculture, 161(1–4), 67–78. Doi:10.1016/S0044-8486(97)00257-3.

Carter C.G., Hauler R.C., 2000: Fish meal replacement by plant meals in extruded feeds for Atlantic salmon, *Salmo salar* L. Aquaculture, 185, 299–311.

Dan Z.J., Zhang W.C., Zheng J.C., Gong Y., Cui K., Mai K., Ai Q.H., 2022. Effects of fishmeal substitution by four fermented soybean meals on growth , antioxidant capacity and inflammation responses of turbot juveniles (*Scophthalmus maximus* L.). Aquaculture, 560.

De Santis C., Tocher D.R., Ruohonen K., El-Mowafi A., Martin S.A.M., Dehler C.E., Secombes C.J., Crampton V., 2016: Air-classified faba bean protein concentrate is efficiently utilized as a dietary protein source by post-smolt Atlantic salmon (*Salmo salar*). Aquaculture, 452, 169–177.

Dersjant-Li Y., 2021: The use of soya protein in aquafeeds. Oilseeds, Focus 7, 29–31.

DLG (Deutsche Landwirtschafts-Gesellschaft), 2014: DLG-Futterwerttabellen Schweine. DLG-Verlag, Frankfurt/Main, 42–43.

Durst L., Freitag M., Bellof G, 2021: Futtermittel für landwirtschaftliche Nutztiere. DLG Verlag Frankfurt/Main, 258–304.

Ettle T., Obermaier A., Rimili S., 2011: Eiweiß für die Milchkuh – Sojaprodukte aus heimischer Aufbereitung in der Milchviehfütterung. Bayerische Landesanstalt für Landwirtschaft Grub. Fachartikel in: www.proteinmarkt.de, September 2011.

Evonik, 2010: AminoDat 4.0. Evonik Industries.

Facciolongo A.M., Rubino G., Zarrilli A., Vicenti A., Ragni M., Toteda, F., 2014: Alternative protein sources in lamb feeding 1. Effects on productive performances, carcass characteristics and energy and protein metabolism. Progress in nutrition, 16, 105–115.

Feedipedia, 2018: Animal feed resources information system – INRA CIRAD AFZ and FAO. https://www.feedipedia.org/.

Fuchs V., Schmidt J., Slater M., Zentek J., Buck B., Steinhagen D., 2015: The effect of supplementation with polysaccharides, nucleotides, acidifiers and Bacillus strains in fish meal and soy bean based diets on growth performance in juvenile turbot (*Scophthalmus maximus*). Aquaculture, 437, 243–251.

Galkanda-Arachchige H.S., Guo J., Stein H.H., Allen Davis D., 2020: Apparent energy, dry matter and amino acid digestibility of differently sourced soybean meal fed to Pacific white shrimp *Litopenaeus vannamei*. Aquaculture research, 51(1), 326–340.

Hardy R.W., Kaushik S.J., 2021: Fish nutrition. Academic press.

Hardy R.W., 2000: New developments in aquatic feed ingredients, and potential of enzyme supplements. Avances en Nutricion Acuicola V. Memorias del V Simposium Internacional de Nutricion Acuicola, 19–22 Noviembre, 2000. Merida, Yucatan, Mexico.

Jeroch H., Flachowsky G., Weissbach F., 1993: Futtermittelkunde. Gustav Fischer Verlag Jena, Stuttgart.

Jeroch H., Lipiec H., Abel H.J., Zentek J., Grela E.R., Bellof G., 2016: Körnerleguminosen als Futter- und Nahrungsmittel. DLG-Verlag, Frankfurt/M., 1. Auflage.

Jiang D.L., Zheng J.C., Dan Z.J., Tang Z.Y., Ai Q.H., Mai K.S., 2019: Effects of five compound attractants in high plant-based diets on feed intake and growth performance of juvenile turbot (*Scophthalmus maximus*). Aquaculture Research 50(9), 2350–2358.

Kader M.A., Bulbul M., Koshio S., Ishikawa M., Yokoyama S., Nguyen B.T., Komilus C.F., 2012: Effect of complete replacement of fishmeal by dehulled soybean meal with crude attractants supplementation in diets for red sea bream. Aquaculture, 350, 109–116. https://doi.org/10.1016/j.aquaculture.2012.04.009.

Kaushik S., Cravedi J.-P., Lalles J., Sumpter J., Fauconneau B., Laroche M., 1995: Partial or total replacement of fish meal by soybean protein on growth, protein utilization, potential estrogenic or antigenic effects, cholesterolemia and flesh quality in rainbow trout, *Oncorhynchus mykiss*. Aquaculture, 133, 257–274.

Kissil G.W., Lupatsch I., Higgs D.A., Hardy R.W., 2000: Dietary substitution of soy and rapeseed protein concentrates for fish meal, and their effects on growth and nutrient utilization in gilthead seabream *Sparus aurata* L. Aquaculture research, 31, 595–601.

Kotzamanis Y., Tsironi T., Brezas A., Grigorakis K., Ilia V., Vatsos I., Romano N., van Eys J., Kumar V., 2020: High taurine supplementation in plant protein-based diets improves growth and organoleptic characteristics of European seabass. Scientific reports 10(1).

Krogdahl Å., Bakke-McKellep A.M., Baeverfjord G., 2003: Effects of graded levels of standard soybean meal on intestinal structure, mucosal enzyme activities, and pancreatic response in Atlantic salmon (*Salmo salar* L.). Aquaculture nutrition, 9, 361–371.

Krogdahl Å., Penn M., Thorsen J., Refstie S., Bakke A.M., 2010: Important antinutrients in plant feedstuffs for aquaculture: an update on recent findings regarding responses in salmonids. Aquaculture research, 41, 333–344.

Lee K.-J., Powell M.S., Barrows F.T., Smiley S., Bechtel P., Hardy R.W., 2010: Evaluation of supplemental fish bone meal made from Alaska seafood processing byproducts and dicalcium phosphate in plant protein based diets for rainbow trout (*Oncorhynchus mykiss*). Aquaculture, 302, 248–255.

LfL Bayern, 2014: Futterberechnung für Schweine. ES-Druck, 85356 Freising-Tüntenhausen. 21. Auflage.

LfL Bayern, 2017: Gruber Tabelle zur Fütterung der Milchkühe Zuchtrinder Schafe Ziegen. Bayerische Landesanstalt für Landwirtschaft, 42. Auflage.

Li Y., Gajardo K., Jaramillo-Torres A., Kortner T.M., Krogdahl A., 2022: Consistent changes in the intestinal microbiota of Atlantic salmon fed insect meal diets. Animal microbiome, 4:8, 2–15.

Losand B., Pries M., Steingaß H., Bellof G., 2020: Praxisinformation „Ackerbohnen, Körnerfuttererbsen, Süßlupinen und Sojabohnen in der Rinderfütterung". Editor: Union zur Förderung von Öl- und Proteinpflanzen e.V. (UFOP), Berlin, 2. aktualis. Auflage.Martin J., 2013: Ackerbohnen – Rohproteinquelle für die Lämmermast. Zeitschrift Schafe aktuell in Mecklenburg-Vorpommern, 18, 27–35.

Mambrini M., Roem A.J., Carvèdi J.P., Lallès J.P., Kaushik S.J., 1999: Effects of replacing fish meal with soy protein concentrate and of DL-methionine supplementation in high-energy, extruded diets on the growth and nutrient utilization of rainbow trout, *Oncorhynchus mykiss*. Journal of animal science, 77, 2990–2999.

McLean E., Barrows F.T., Craig S.R., Alfrey K., Tran L., 2020: Complete replacement of fishmeal by soybean and poultry meals in Pacific whiteleg shrimp feeds: Growth and tolerance to EMS/AHPND and WSSV challenge. Aquaculture, 527, 735383.

Metochis C.P., Crampton V.O., Ruohonen K., Bell J.G., Adams A., Thompson K.D., 2016a: The effects of increasing dietary levels of amino acid-supplemented soy protein concentrate and constant dietary supplementation of phosphorus on growth, composition and immune responses of juvenile Atlantic salmon (*Salmo salar* L.). Fish physiology and biochemistry, 42, 807–829.

Metochis C.P., Spanos I., Auchinachie N., Crampton V.O., Bell J.G., Adams A., Thompson K.D., 2016b: The effects of increasing dietary levels of soy protein concentrate (SPC) on the immune responses and disease resistance (furunculosis) of vaccinated and non-vaccinated Atlantic salmon (*Salmo salar* L.) parr. Fish & shellfish immunology, 59, 83–94.

Mogensen L., Lund P., Kristensen T., Weisbjerg M.R., 2008: Effects of toasting blue lupins, soybeans or barley as supplement for high-yielding, organic dairy cows fed grass-clover silage ad libitum. Livestock science, 115, 249–257.

Mosenthin R., Jezierny D., Eklund M., 2004: Untersuchungen zur Bestimmung der standardisierten praecaecalen Verdaulichkeiten von Protein und Aminosäuren aus Körnerleguminosen beim Schwein. Editor: Union zur Förderung von Oel- und Proteinpflanzen e.V. (UFOP), 2004: Abschlussbericht zum UFOP-Projekt. Berlin.

Nengas I., Alexis M.N., Davies S.J., 1996: Partial substitution of fishmeal with soybean meal products and derivatives in diets for the gilthead sea bream *Sparus aurata*. Aquaculture research, 27, 147–156.

Naumann C., Bassler R., 1988: 2. Ergänzungslieferung in VDLUFA-Methodenbuch, Vol. III. Die chemische Untersuchung von Futtermitteln. VDLUFA-Verlag, Darmstadt.

Olli J.J., Krogdahl Å., van den Ingh T.S.G.A.M. and Brattås L.E., 1994: Nutritive value of four soybean products in diets for Atlantic salmon (*Salmo salar*, L.). Acta agriculturae Scandinavica, section A – animal science, 44, 50–60.

Peng M., Xu W., Ai Q., Mai K., Liufu Z., Zhang K., 2013: Effects of nucleotide supplementation on growth, immune responses and intestinal morphology in juvenile turbot fed diets with graded levels of soybean meal (*Scophthalmus maximus* L.). Aquaculture, 392, 51–58.

Petereit J., Hoerterer C., Bischoff-Lang A.A., Conceição L.E.C., Pereira G., Johansen J., Pastres R., Buck B.H., 2022: Adult European sea bass (*Dicentrarchus labrax*) perform well on alternative circular-economy-driven feed formulations. Sustainability, 14, 7279.

Puntigam R., Slama J., Rittler l., 2021: Studie zur Sojaverarbeitung 2020. Ergebnisbericht, Teil 1. Editor: Donau Soja, Wien.

Rahimnejad S., Lu K.L., Wang L., Song K., Mai K.S., Davis D.A., Zhang C.X., 2019: Replacement of fish meal with SE5 and ZR1 fermented soybean meal in diets for Japanese seabass. Fish & shellfish immunology, 84, 987–997. Doi:10.1016/j.fsi.2018.11.009.

Ray G.W., Liang D., Yang Q., Tan B., Dong X., Chi S., Liu H., Zhang S., Rimei, L., 2020: Effects of replacing fishmeal with dietary soybean protein concentrate (SPC) on growth, serum biochemical indices, and antioxidative functions for juvenile shrimp *Litopenaeus vannamei*. Aquaculture, 516, 734630.

Refstie S., Sahlström S., Bråthen E., Baeverfjord G., Krogedal P., 2005: Lactic acid fermentation eliminates indigestible carbohydrates and antinutritional factors in soybean meal for Atlantic salmon (*Salmo salar*). Aquaculture, 246, 331–345.

Refstie S., Storebakken T., Roem A.J., 1998: Feed consumption and conversion in Atlantic salmon (*Salmo salar*) fed diets with fish meal, extracted soybean meal or soybean meal with reduced content of oligosaccharides, trypsin inhibitors, lectins and soya antigens. Aquaculture, 162, 301–312.

Robaina L., Izquierdo M., Moyano F., Socorro J., Vergara J., Montero D., Fernandez-Palacios H., 1995: Soybean and lupin seed meals as protein sources in diets for gilthead seabream (*Sparus aurata*): nutritional and histological implications. Aquaculture, 130, 219–233.

Saitoh S., Koshio S., Harada H., Watanabe K., Yoshida T., Teshima S.-I., Ishikawa, M., 2003: Utilization of extruded soybean meal for Japanese flounder *Paralichthys olivaceus* juveniles. Fisheries science, 69(5), 1075–1077.

Sauvant D., Perez J.-M., Tran G., 2004: Tables of composition and nutritional value of feed materials. INRA, 2. Auflage.

Shiu Y.L., Wong S.L., Guei W.C., Shin Y.C., Liu C.H., 2015: Increase in the plant protein ratio in the diet of white shrimp, *Litopenaeus vannamei* (Boone), using *Bacillus subtilis* E20-fermented soybean meal as a replacement. Aquaculture research, 46, 382–394.

Song Z., Li H., Wang J., Li P., Sun Y., Zhang, L., 2014: Effects of fishmeal replacement with soy protein hydrolysates on growth performance, blood biochemistry, gastrointestinal digestion and muscle composition of juvenile starry flounder (*Platichthys stellatus*). Aquaculture, 426, 96–104.

Steiner T., Bellof G., 2009: Einsatz von unterschiedlich wärmebehandelten Sojakuchen in der ökologischen Hähnchenmast.10. Wissenschaftstagung Ökologischer Landbau, 11.–13.2.2009, ETH Zürich, Tagungsband.

Stickney R.R., Hardy R.W., Koch K., Harrold R., Seawright D., Massee K.C., 1996: The effects of substituting selected oilseed protein concentrates for fish meal in rainbow trout *Oncorhynchus mykiss* diets. Journal of the world aquaculture society, 27, 57–63.

Storebakken T., Shearer K.D., Roem A.J., 1998: Availability of protein, phosphorus and other elements in fish meal, soy-protein concentrate and phytase-treated soy-protein-concentrate-based diets to Atlantic salmon, *Salmo salar*. Aquaculture, 161, 365–379.

Storebakken T., Shearer K.D., Roem A.J., 2000: Growth, uptake and retention of nitrogen and phosphorus, and absorption of other minerals in Atlantic salmon *Salmo salar* fed diets with fish meal and soy–protein concentrate as the main sources of protein. Aquaculture Nutrition, 6, 103–108.

Takagi S., Shimeno S., Hosokawa H., Ukawa M., 2001: Effect of lysine and methionine supplementation to a soy protein concentrate diet for red sea bream. Fisheries science, 67, 1088–1096.

Thurner S., 2020: Optimierung dezentraler Sojaaufbereitungsverfahren mittels Online-Prozesssteuerung über Nahinfrarot-Spektroskopie. Schlussbericht BÖLN-Projekt (FKZ 2814EPS022).

Tibaldi E., Hakim Y., Uni Z., Tulli F., de Francesco M., Luzzana U., Harpaz, S., 2006: Effects of the partial substitution of dietary fish meal by differently processed soybean meals on growth performance, nutrient digestibility and activity of intestinal brush border enzymes in the European sea bass. Aquaculture, 261(1), 182–193. Doi:10.1016/j.aquaculture.2006.06.026.

Urán P., Schrama J., Rombout J., Obach A., Jensen L., Koppe W., Verreth J., 2008: Soybean meal-induced enteritis in Atlantic salmon (*Salmo salar* L.) at different temperatures. Aquaculture nutrition, 14, 324–330.

Venou B., Alexis M.N., Fountoulaki E., Haralabous J., 2006: Effects of extrusion and inclusion level of soybean meal on diet digestibility, performance and nutrient utilization of gilthead sea bream. Aquaculture, 261, 343–356.

Vielma J., Mäkinen T., Ekholm P., Koskela J., 2000: Influence of dietary soy and phytase levels on performance and body composition of large rainbow trout (*Oncorhynchus mykiss*) and algal availability of phosphorus load. Aquaculture, 183, 349–362.

Watson A.M., Barrows F.T., Place A.R., 2013: Taurine supplemented plant protein based diets with alternative lipid sources for juvenile gilthead sea bream, *Sparus aurata*. Journal of fisheries and aquaculture, 4, 59–66.

Weber M., 2018: Körnerleguminosen – 3 Jahre bundesweites Monitoring der Inhaltsstoffe. Veredlungsproduktion. https://www.proteinmarkt.de/aktuelles/archiv/details/news/koernerleguminosen-3-jahre-bundesweites-monitoring-der-inhaltsstoffe. Retrieved: August 7, 2024.

Weber M., Preissinger W., Bellof G., 2020: Praxisinformation „Ackerbohnen, Körnerfuttererbsen, Süßlupinen und Sojabohnen in der Schweinefütterung“. Editor: Union zur Förderung von Öl- und Proteinpflanzen e.V. (UFOP), Berlin, 2. aktualis. Auflage.

Yamamoto T., Iwashita Y., Matsunari H., Sugita T., Furuita H., Akimoto A., Okamatsu K., Suzuki N., 2010: Influence of fermentation conditions for soybean meal in a non-fish meal diet on the growth performance and physiological condition of rainbow trout *Oncorhynchus mykiss*. Aquaculture, 309, 173–180.

Yang Y.-H., Wang Y.-Y., Lu Y., Li Q.-Z., 2011: Effect of replacing fish meal with soybean meal on growth, feed utilization and nitrogen and phosphorus excretion on rainbow trout (*Oncorhynchus mykiss*). Aquaculture international, 19, 405–419.

Zhang Y., Yang P.P., Sun H.R., Hou Y.Q., Zhang Y., Liu H.Y., 2023: Evaluation of extruded full-fat soybean as the substitution for fish meal in diets for juvenile *Scophthalmus maximus* based on growth performance, intestinal microbiota, and aquaculture water quality. Aquaculture, 562.

2.8 Other grain legumes

Gerhard Bellof

In addition to the grain legumes presented in detail, other members of this genus can be used as animal feed. The relevant ingredients, feed value and feeding recommendations for lentils (*Lens culinaris L.*), chickpeas (*Cicer arietinum L.*) and vetches (*Vicia sativa L.*) are presented below.

FIGURE 2.8-1 Vetches (flowering plants)
PHOTO © K.-P. WILBOIS

2.8.1 *Ingredients of lentils (Lens culinaris L.), chickpeas (Cicer arietinum L.) and vetches (Vicia sativa L.)*

Table 2.8.1-1 shows the ingredients of lentils, chickpeas and vetches (fodder vetches). These grain legumes have a crude protein content that lies between that of peas and field beans.

Similar to these, the above-mentioned grain legumes contain high levels of starch. Vetches have a similarly high starch content to peas; lentils are slightly lower. With 313 g starch/kg (at 88% DM), chickpeas rank third in this list and are therefore lower than the starch content of field beans.

The crude fat content of lentils, chickpeas and vetches is at a low level. However, the crude fat of lentils, chickpeas and peas has a favourable ratio of polyunsaturated fatty acids from the omega-n fatty acid family (n-6/n-3).

Of the three grain legumes mentioned, chickpeas have the highest crude fibre and ADF content (121 g/kg, at 88% DM). They thus almost reach the level of sweet lupins (180 g ADF/kg).

Lentils, chickpeas and vetches have low calcium contents (1.1 g–1.7 g/kg DM) and medium phosphorus contents (3.4 g–4.5 g/kg DM). This makes them comparable to peas. While the potassium content of vetches (11.1 g/kg DM) and chickpeas (10.5 g/kg DM) is similar to that of field beans, lentils have significantly lower values (4.5 g/kg DM). The ranking for sodium content is comparable – at a low level.

The trace element contents are in the range of the already treated grain legumes.

TABLE 2.8.1-1 Ingredients (g/kg) of lentils, chickpeas and vetches (fodder vetches)

Ingredient	Lentils		Chickpeas		Vetches	
DM	880	1000	880	1000	880	1000
CP	237	269	194	221	250	284
CL	14	16	41	47	13	15
CF	43	49	92	105	41	47
aNDFom	n.d.	130	201	228	138	157
ADFom	n.d.	63	121	138	66	75
STA	402	457	313	356	429	487
SUG	0	0	32	36	16	18
NFE	553	628	523	594	540	614
ASH	33	38	29	33	35	40

SAUVANT *ET AL.* (2004); JEROCH *ET AL.* (2016); HUANG *ET AL.* (2017); FEEDIPEDIA (2018)

FIGURE 2.8.1-1
Vetches (grains)
PHOTO © K.-P. WILBOIS

2.8.2 *Feed value of lentils, chickpeas and vetches*

2.8.2.1 Energy

The feeding of lentils, chickpeas or vetches is most likely to be considered for monogastric livestock. Due to their high protein and starch content, lentils and chickpeas have a high ME (or NE) content for pigs. This means that lentils reach the level of peas and vetches just under the level of field beans. Chickpeas, on the other hand, are significantly lower in ME content (pigs) due to their lower starch content and comparatively high crude fibre values.

Chickpeas also provide the lowest AME_N content for poultry and can therefore be equated with sweet lupins. Vetches, on the other hand, reach the AME_N level of peas. Lentils occupy a middle position here.

2.8.2.2 Protein

Table 2.8.2-1 summarises the amino acid contents (g/kg) of lentils, chickpeas and vetches. The amino acid profiles of lentils and chickpeas shown there are similar to those of peas. Consequently, there is a deficit of sulphur-containing amino acids and tryptophan when these legumes are used in pig and poultry feed. The amino acid pattern of seed vetches differs markedly from the grain legumes mentioned. In particular, the levels of methionine and tryptophan are almost twice as high as in the comparable feedstuffs. They almost reach the level of soybeans. The lysine content is also high in common vetch. It exceeds the corresponding content value of the pea and almost reaches the level of the field bean. This makes hairy vetch an interesting protein feed, especially for organic pig and poultry feed.

TABLE 2.8.2-1 Amino acid contents (g/kg) of lentils, chickpeas and vetches (fodder vetches)

DM/ Amino acid	Lentils		Chickpeas		Vetches	
DM	880	1000	880	1000	880	1000
Lys	15.4	17.5	12.8	14.6	16.7	19.0
Met	2.1	2.4	2.3	2.7	4.2	4.8
Cys	2.8	3.2	2.3	2.7	1.7	2.0
Thr	8.3	9.4	6.6	7.5	8.0	9.1
Try	1.9	2.2	1.8	2.0	4.0	4.5
Ile	10.7	12.1	7.4	8.4	9.5	10.8
Leu	18.0	20.4	13.8	15.7	17.7	20.2
Val	10.9	12.4	7.6	8.6	10.5	11.9

JEROCH *ET AL.* (2016A) HUANG *ET AL.* (2017); FEEDIPEDIA (2018)

2.8.3 *Feeding recommendations for lentils, chickpeas and vetches*

2.8.3.1 Pigs

In complete feed mixtures for rearing piglets (weaned), no negative influence on feed intake was found with a proportion of 30% lentils. However, crude protein digestibility and daily weight gain were reduced (Landero *et al.*, 2012). The authors therefore recommend 20% lentils as the maximum proportion in complete feed mixtures for rearing piglets.

Raw chickpeas have medium to high levels of antinutritive substances (protease inhibitors, tannins), which can impair intestinal nutrient absorption in pigs.

Nevertheless, up to 50% chickpea content is possible in complete feed mixtures for pigs in exchange for SES (Jeroch *et al.*, 2016).

Vetches contain the hydrocyanic acid glycoside vicianin, which is hydrolysed to glucose and hydrocyanic acid (HCN) in the intestinal tract of animals. Individual vetch varieties can reach high levels of vicianin (e.g. up to 430 mg/100g), which release highly toxic amounts of hydrocyanic acid in these concentrations (Jeroch *et al.*, 2016). In a feeding trial with weaned piglets, Seabra *et al.* (2001) used a 25% proportion of vetches in the rearing mixtures. They found no difference in feed intake and weight development compared to the control. The analysed digestibility of nutrients was also the same for both groups. Collins *et al.* (2005) tested vetch proportions of 7.5%, 15% and 22.5% in complete feed mixtures in a feeding trial with fattening pigs. Up to a proportion of 15%, no negative effects on feed intake, fattening performance and carcass value were observed.

TABLE 2.8.3-1 Recommendations for the use of lentils, chickpeas and vetches in pig feed (maximum mixture proportions for complete feed mixtures, figures in %)

Animal group	Lentils	Chickpeas	Vetches
Piglets			
≤ 20 kg	10	10	15
> 20 kg	20	10	15
Fattening pigs			
Starter	20	20	15
Finisher	20	30	20
Sows[2]			
Pregnant	5	6	5
Lactate	(15)[1]	(15)[1]	(20)[1]

1 Preliminary studies.

JEROCH *ET AL.* (2016); HUANG *ET AL.* (2017), MODIFIED

2.8.3.2 Poultry

Only a few poultry studies for lentils and chickpeas can be found in the literature. In contrast, more recent results are available for the use of seed vetches in poultry.

Farran *et al.* (1995) tested increasing proportions of vetches (7.5%, 15.0% and 22.5%) in AF for laying hens. The hens only reacted with a decrease in feed intake, reduced laying performance and increased feed consumption (kg feed/kg egg mass) from a mixture proportion of 22.5%. However, the performance level was relatively low. In a follow-up experiment, Farran *et al.* (2001) investigated the effect of different treatments of vetches for laying hens. The vetches were treated either in a water bath (40 °C) (V1) or an acetic acid bath (40 °C) (V2). Ratios of 60% vetches were used in the complete feed mixtures. The animals in group V1 showed no significant drop in performance compared to the control group, while those in V2 showed increased feed consumption (kg feed/kg egg mass). Kaya *et al.* (2013) also tested the treatment effect of vetches. They carried out either an autoclaving or a roasting treatment. These variants as well as untreated vetches were used in ratios of 25% in complete feed mixtures for laying hens. All test groups showed reduced feed intake and a significant drop in laying performance compared to the control.

There are some recent studies on the use of vetches for broiler fattening. Sadeghi *et al.* (2011) tested the use of untreated and heat-treated vetches (10%, 20% and 30% respectively) in complete feed mixtures for broilers (22nd to 42nd day of life). Although the daily feed intake in the experimental groups remained high compared to the control group, the daily weight gain and feed consumption per kg gain showed a clear negative dose effect for both the groups with untreated and treated vetches.

TABLE 2.8.3-2 Recommendations for the use of seed vetches in poultry feed (maximum mixture proportions for complete feed mixtures, figures in %)

Animal group	Vetches[1]
Laying hens	
Egg production	15
Reproduction	15
Broiler	
Starter (≤. 4. LW)[2]	10
Finisher (> 4. LW)	15

1 Untreated.
2 LW: week of life.
JEROCH *ET AL.* (2016); HUANG *ET AL.* (2017), MODIFIED

Saki *et al.* (2008) came to similar results in an analogue experimental design (male broilers; phase 1: 1-21, phase 2: 22-42 day of life; untreated vs. heat-treated vetches; 10%, 20% and 30% mixture content respectively). However, the two groups with 30% vetches in the complete feed mixtures already showed a significant drop in feed intake and growth depression in the first phase.

Table 2.8.3-2 shows the recommendations for the use of seed vetches in poultry feed.

2.8.3.3 Fishes and shrimps

Monika Weiß, Sinem Zeytin and Vanessa Fuchs

Knowledge about the use of other grain legumes in fish feed is still scarce, however a few feeding trials have been conducted with various plant species, including Narbonne vetch (*Vicia narbonensis L.*), common vetch (*Vicia sativa L.*) in the diets. A overview of studies testing vetch ingredients in aquafeeds for various species is given in table 2.8.3-3.

In feeds for rainbow trout, seed meal of Narbonne vetch is applicable at levels up to 10% of the feed. Higher inclusion rates lead to an impairment of the activity of digestive enzymes, leading to biochemical alterations which ultimately reduce growth. These alterations are possibly related to the presence on non-starch polysaccharides (NSP) in Narbonne vetch. Although further research is needed, Narbonne vetch is a candidate to enhance sustainability of European Aquafeeds in reducing its carbon footprint (Tomás-Almenar *et al.*, 2020; Toledo-Solís *et al.*, 2023).

TABLE 2.8.3-3 Executive summary of studies testing vetch ingredients in aquafeeds for various species

Green legume	Target species	Initial mean weight (g)	Product	Replaced resource	Inclusion rate (%)	Recommendable inclusion rate	Notes	Reference
Narbonne vetch (*Vicia narbonensis*)	Rainbow trout (*Oncorhynchus mykiss*)	~38	Seed meal	SBM	10, 20		Study identifies future strategies to improve the use of Narbonne vetch meal in Aquaculture	Toledo-Soliz *et al.* 2023
	Rainbow trout (*Oncorhynchus mykiss*)	~27	Ground and sieved (0.7 mm)	Reference diet	10, 30	10	Might reduce carbon footprint in European aquaculture and the dependency on imported resources	Tomás-Almenar *et al.* 2020
	Nile tilapia (*Oreochromis niloticus*)	~12	Seeds	Reference diet	0.1, 0.2, 0.3 + 0.4	20		Buyukcapar *et al.* 2010
Common vetch (*Vicia sativa*)	Australian silver perch (*Bidyanus bidyanus*)	~16	Whole and dehulled	Reference diet	30	Not	Poorly accepted	Booth *et al.* 2001

For Nile tilapia (*Oreochromis niloticus*) acceptable inclusion rates are higher, but should not exceed 20% (Buyukcapar *et al.*, 2010).

Common vetch *Vicia sativa* was tested whole and dehulled in feeds for the Australian silver perch (*Bidyanus bidyanus*) (Allan *et al.*, 2000; Booth *et al.*, 2001) at an inclusion rate of 30%. Although dehulling improved the nutritional value immensely, diets where poorly accepted, which indicates that the inclusion of *Vicia sativa* is not recommendable for the Australian silver perch, at least not in comparable high inclusion rates. It was suggested that the refusal of the diets could be linked to toxic amino acids in this plant species.

References

Allan G.L., Rowland S.J., Mifsud C., Glendenning D., Stone D.A.J., Ford A., 2000: Replacement of fish meal in diets for Australian silver perch, *Bidyanus bidyanus*: V. Least-cost formulation of practical diets. Aquaculture, 186, 327–340.

Booth M.A., Allan G.L., Frances J., Parkinson S., 2001: Replacement of fish meal in diets for Australian silver perch, *Bidyanus bidyanus*: IV. Effects of dehulling and protein concentration on digestibility of grain legumes. Aquaculture, 196, 67–85.

Buyukcapar H., Mezdegi M., Kamalak A., 2010: Nutritive value of narbon bean (*Vicia narbonensis*) seed as ingredient in practical diet for tilapia (*Oreochromis niloticus*) fingerlings. Journal of applied animal research, 37, 253–256.

Collins C.L., Dunshea F.R., Henman D.J., King R.H., 2005: Evaluation of common vetch (*Vicia sativa* cv. Morava) for growing pigs. Austr. J. Exp. Agric., 45, 699–703.

Farran M.T., Dakessian P.B., Darwish A.H., Uwayjan M.G., Dbouk H.K., Sleiman F.T., Ashkarian V.M., 2001: Performance of broilers and production and egg quality parameters of laying hens fed 60% raw or treated common vetch (*Vicia sativa*) seeds. Poultry science, 80, 203–208.

Farran M.T., Uwayjan M.G., Miski A.M.A., Sleiman F.T., Adada F.A., Ashkarian V.M., Thomas O.P., 1995: Effect of feeding raw and treated common vetch seed (Vicia sativa) on the performance and egg quality parameters of laying hens. Poultry science, 74, 1630–1635.

Feedipedia, 2018: Animal feed resources information system – INRA CIRAD AFZ and FAO. https://www.feedipedia.org/.

Huang Y.F., Gao X.L., Nan Z.B., Zhang Z.X., 2017: Potential value of the common vetch (*Vicia sativa* L.) as an animal feedstuff: a review. Journal of animal physiology and animal nutrition, 101, 807–823.

Jeroch H., Lipiec H., Abel H.J., Zentek J., Grela E.R., Bellof G., 2016: Körnerleguminosen als Futter- und Nahrungsmittel. DLG-Verlag, Frankfurt/M., 1. Auflage.

Kaya A., Yeoreuk A.M., Esenburga N., Temelli A., Ekinci E.O., 2013: The effect of raw and processed common vetch seed (*Vicia sativa*) added to diets of laying hens on performance, egg quality, blood parameters and liver histopathology. Journal of poultry science, 50, 228–236.

Landero J.L., Beltranena E., Zijlstra R.T., 2012: The effect of feeding lentil on growth performance and diet nutrient digestibility in starter pigs. Anim. Feed sci. Technol., 174, 108–112.

Sadeghi G.H., Tabeidian S.A., Toghyani M., 2011: Effect of processing on the nutritional value of common vetch (*Vicia sativa*) seed as a feed ingredient for broilers. Journal of applied poultry research, 20, 498–505.

Saki A.A., Pourhesabi G., Yaghobfar A., Mosavi M.A., Tabatabai M.M., Abbasinezhad M., 2008: Effect of different levels of the raw and processed vetch seed (*Vicia sativa*) on broiler performance. Journal of biological sciences, 8, 663–666.

Sauvant D., Perez J.-M., Tran G., 2004: Tables of composition and nutritional value of feed materials. INRA, 2. Auflage.

Seabra M., Carvalho S., Freire J., Ferreira R., Mourato M., Cunha L., Cabral F., Teixeira F., Aumaitre, A., 2001: *Lupinus luteus, Vicia sativa and Lathyrus cicera* as protein sources for piglets: ileal and total tract apparent digestibility of amino acids and antigenic effects. Animal feed science and technology, 89, 1–16.

Toledo-Solís F.J., Larrán A.M., Martín B., de la Cuesta P.L., Mateos-Aparicio I., Pérez V., Moyano F.J., Fernández I., 2023: Uncovering the physiological impacts of soybean meal replacement by Narbonne vetch (*Vicia narbonensis*) meal in rainbow trout (*Oncorhynchus mykiss*) diets: towards the future and sustainable European aquaculture. Animal feed science and technology, 296, 115555.

Tomás-Almenar C., Toledo-Solís F.J., Larrán A.M., de Mercado E., Alarcón F.J., Rico D., Martín-Diana A.B., Fernández I., 2020: Effects and safe inclusion of narbonne vetch (*Vicia narbonensis*) in rainbow trout (*Oncorhynchus mykiss*) diets: towards a more sustainable aquaculture. Animals, 10, 2175.

2.9 Feed from grain legume processing

2.9.1 *Protein concentrates and isolates*

Mechthild Freitag

Grain legume isolates with a crude protein content up to 90% are the most purified commercial plant proteins. The degree of purification can be estimated by the protein content of the ingredient and depends on the manufacturing process. The process minimizes the flavour and the colour of the original seed. Protein isolates are prepared from cleaned, dehulled seeds by removing most of the non-protein components such as carbohydrates.

These proteins are processed into several forms like spray-dried powder and dry granules. They are a highly digestible protein source with low levels of antinutritive factors (ANF). Solubility of the isolate and thus the suitability for liquid feed varies according to the production procedure.

2.9.1.1 Soybean (*Glycine max L.*) protein products

In addition to their high nutritional value, soya protein products have a wide range of functional properties. They are used as diet components in the production of a wide variety of feeds for various animals. The most important protein products are soy protein concentrate (SPC, CP ≥ 70%) and isolate (SPI, CP ≥ 90%). These are produced from deoiled soy flakes or dehulled soybeans (Figures 2.9.1-1 and 2.9.1-2).

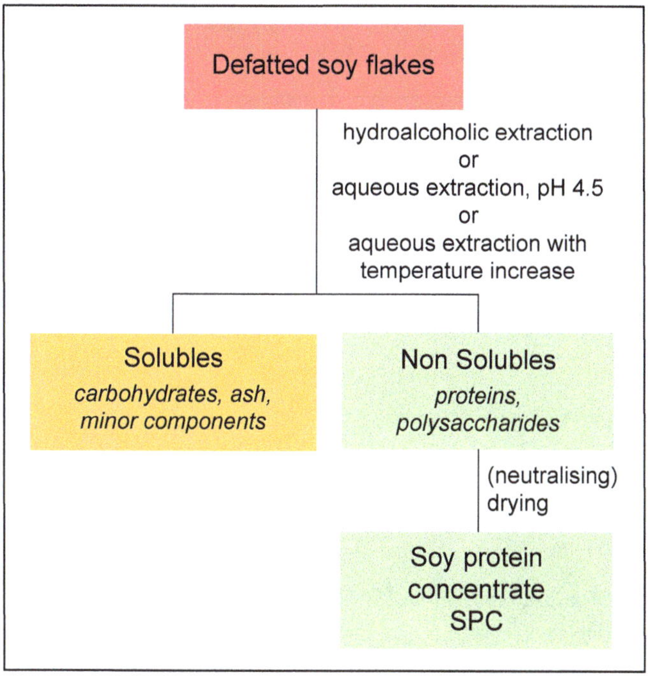

FIGURE 2.9.1-1 Production processes of soy protein concentrate DAZERT, 2004, MODIFIED

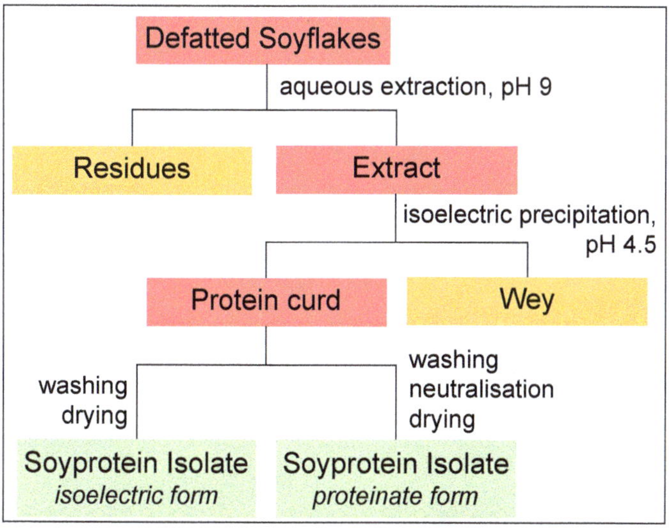

FIGURE 2.9.1-2 Production process of soy protein isolate DAZERT (2004); MODIFIED

2.9.1.1.1 *Ingredients*

SPC contain at least 70% protein in DM (table 2.9.1-1). The proteins are concentrated by extraction in aqueous alcohol or with a dilute acid solution in the pH range of 4.0 to 4.8 (figure 2.9.1-1). During this process polysaccharides and oligosaccharides are partly removed. This makes it a very valuable feed for young animals because, in addition to its high digestibility, it has a very low content of antinutritive

TABLE 2.9.1-1 Ingredients in soy protein concentrate (SPC) and isolate (SPI), compared to soybean meal (SBM) (g/kg; NRC, 1998)

Ingredient	SBM		SPC		SPI	
DM	880	1000	880	1000	880	1000
CP	475	540	635	722	821	933
CL	10	11	6	7	5	6
CF	35	40	32	36	10	11
NDF	89	101	*	*	*	*
ADF	54	61	*	*	*	*
ASH	60	68	51	58	49	55

* No data.
ADM (2024); MODIFIED

TABLE 2.9.1-2 Mineral content of soybean concentrate (SPC) and isolate (SPI), compared to soybean meal (SBM) (NRC, 1998)

Ingredient		SBM		SPC		SPI	
DM	%	88	100	88	100	88	100
ASH	%	6.0	6.8	5.1	5.8	4.9	5.5
Ca	%	0.34	0.38	0.31	0.39	0.14	0.16
P	%	0.69	0.78	0.78	0.89	0.62	0.71
K	%	2.14	2.43	2.15	2.44	0.26	0.29
Mg	%	0.30	0.34	0.31	0.36	0.08	0.09
S	%	0.44	0.50	*	*	0.68	0.77
Cu	ppm	20.0	22.7	12.7	14.4	13.4	15.2
Fe	ppm	176.0	200	107.5	122.2	131.0	148.9
Mn	ppm	36.0	40.9	*	*	4.8	5.4
Zn	ppm	55.0	62.5	29.3	33.3	32.5	37.0

* No data.
ADM (2024); MODIFIED

factors (ANF) such as trypsin inhibitors, phytate and antigenic ingredients such as glycinin and beta-conglycinin.

SPI are produced by extracting the soy flour with a dilute alkali (pH < 9) followed by centrifugation. The extract is precipitated at pH 4.5. The acid-precipitated protein curd is then centrifuged, washed, neutralized, and spray-dried to produce SPI (figure 2.9.1-2). SPI contains at least 90% protein in DM. It is almost free of carbohydrates and has only very low oil residues. Minerals, on the other hand, are only reduced to a small extent.

During the production process from SBM to SPC the mineral content is slightly altered. The amount of bulk elements is fairly stable, but trace elements, like Fe, Mn and Zn are greatly reduced (table 2.9.1.1-2).

2.9.1.1.2 *Feed value*

Variations in protein quality among soybean products are due to the protein and amino acid (AA) level and the bioavailability of the AAs in the product. The AA profile of soya protein complements the AA profile of cereal grains due to the relatively high concentrations of lysine and tryptophan.

TABLE 2.9.1-3 Comparison of protein content and amino acid composition (% in DM) of various soybean products (NRC 1998)

Amino acid	Full-fat soybeans	Soybean meal		Soy protein concentrate	Soy protein isolate
		Hulled	Dehulled		
Crude protein	35.20	43.80	47.50	64.00	85.80
Lysine	2.22	2.83	3.02	4.20	5.26
Methionine	0.53	0.61	0.67	0.90	1.01
Cysteine	0.55	0.70	0.74	1.00	1.19
Threonine	1.41	1.73	1.85	2.80	3.17
Tryptophan	0.48	0.61	0.65	0.90	1.08
Arginine	2.60	3.23	3.48	5.79	6.87
Histidine	0.96	1.17	1.28	1.80	2.25
Isoleucine	1.61	1.99	2.16	3.30	4.25
Leucine	2.57	3.42	3.66	5.30	6.64
Phenylalanine	1.83	2.18	2.39	3.40	4.34
Tyrosine	1.41	1.73	1.85	2.80	3.17
Valine	1.68	2.06	2.27	3.40	4.21

STEIN *ET AL.* (2008)

TABLE 2.9.1-4 Crude protein content (%) and amino acid digestibility for pigs (%) of various soybean products (NRC 1998)

Crude protein/ Amino acid	Full-fat soybeans	Soybean meal		Soy protein concentrate	Soy protein isolate
		Hulled	Dehulled		
Crude protein	35.2	43.8	47.5	64.0	85.8
Lysine	86	89	90	95	91
Methionine	85	91	91	94	92
Cysteine	80	84	87	94	82
Threonine	83	85	87	94	85
Tryptophan	82	87	90	93	88
Arginine	93	93	84	99	99
Histidine	88	90	91	97	91
Isoleucine	84	88	89	95	90
Leucine	86	88	89	95	89
Phenylalanine	88	88	89	97	92
Tyrosine	87	90	90	96	91
Valine	83	86	88	94	89

STEIN *ET AL.* (2008)

In general, the more processed the soy protein is, the greater is the digestibility of the AAs. Fibre and oligosaccharides reduce digestibility. Therefore, AAs in SPC and SPI are usually more digestible than AAs in SBM, and AAs in dehulled soybean meal have a greater digestibility than AAs in non-dehulled soybean meal (table 2.9.1-4). However, processing procedures like heat treatment or drying may reduce digestibility, as well as a high feed intake. The variation in AA bioavailability among soybean products is primarily due to either insufficient or excessive heat processing. Therefore, AAs in SPC usually have a higher digestibility than those of SPI.

According to ADM (2024), the metabolisable energy values of SPC and SPI for pigs and poultry are both slightly higher than that of corn. However, due to the high protein and virtually no carbohydrate content, the net energy value is more than 15% lower.

2.9.1.1.3 *Feeding recommendations*

SPC and SPI are a better ingredient than SBM in diets fed to weaned piglets because they cause less antigenic reactions in the small intestine. During the production process ANFs are reduced or – in the case of SPI – even eliminated. Therefore, these high-protein soy products are well suited for young animal diets. Because of the SPI's excellent water dispersion characteristics, it is especially suited for milk replacers. However, in many farming situations the high price of both will limit their inclusion in the diet.

2.9.1.2 Grain pea (*Pisum sativum L.*) and faba bean (*Vicia faba L.*) protein products

Grain pea and faba bean protein ingredients can be obtained through wet extraction, dry fractionation or mild extraction (figure 2.9.1-3).

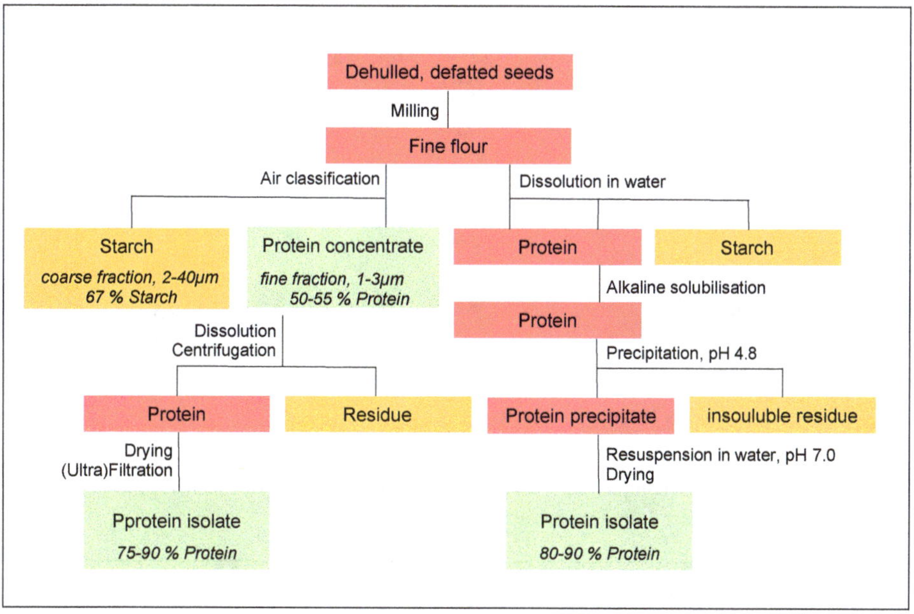

FIGURE 2.9.1-3 Sample production process for grain pea and faba bean protein concentrates and isolates
BOUKID *ET AL.* (2021); MODIFIED

2.9.1.2.1 *Nutrient composition*

Commercial pea protein ingredients include flour (20–25% CP), concentrate (50–75% CP), and isolate (>80% CP). Typically, pea protein concentrates (PPC) contain about 55% protein, 2% lipids, and 35% carbohydrates, e.g. starch and fibre compounds. Faba bean concentrates are higher in protein and lower in oligosaccharide and sugar content (table 2.9.1-5).

Both protein isolates contain about 80–95% protein. They have a very low carbohydrate content and therefore do not cause allergic reactions, which can be the case with protein concentrates. However, as demonstrated in table 2.9.1.2-2, chemical composition and properties of legume isolates vary according to processing procedures (Sumner *et al.*, 1981). These seem to be more important than the plant species the product is derived from. This is most relevant for water absorption characteristics and therefore for the use in milk replacers.

TABLE 2.9.1-5 Chemical composition of pea protein and of faba bean protein concentrate

Ingredient		Pea protein concentrate		Faba bean protein concentrate	
DM	g/kg	880	1000	880	1000
ASH	g/kg	48	55	51	58
CP	g/kg	521	592	621	706
CL	g/kg	23	26	20	23
Carbohydrates					
CF	g/kg	22	25	21	24
STA	g/kg	62	70	38	43
Glucose	%	0.27	0.31	0.35	0.40
Sucrose	%	1.81	2.06	1.46	1.66
Raffinose	%	0.93	1.06	0.35	0.40
Stachyose	%	3.07	3.49	1.31	1.49
Verbascose	%	3.64	4.17	2.79	3.17
Total oligosaccharides	%	7.67	8.72	4.45	5.06
Total sugars	%	9.76	11.09	6.27	7.12

BHATTY AND CHRISTISON (1984); MODIFIED

TABLE 2.9.1-6 Chemical composition (%) and properties (%) of pea (PPI) and soy protein isolates (SPI) after different processing procedures

Ingredient	Na propionate extracted PPI		Isoelectric precipitation PPI		Na-propionate extracted SPI	Isoelectric precipitation SPI
	Spray dried	Freeze dried	Spray dried	Freeze dried	Spray dried	Spray dried
Drying pH	4.5	4.5	7	7	4.5	7
Moisture	5.0	2.4	8.5	4.2	3.8	5.9
CP	85.8	83.0	88.5	90.0	93.4	89.0
CL	5.3	5.4	5.3	4.4	0.1	0.2
CF	0.5	0.7	0.2	0.2	0.2	0.2
ASH	5.2	5.0	2.7	2.8	4.0	3.7
NFE	3.2	6.8	3.3	2.8	2.3	6.9
N solubility, pH 4.5	1	3	3	2	3	0
N solubility, pH 7	63	64	72	87	61	30
Water absorption	250	205	132	112	698	155

SUMNER *ET AL.* (1981)

When feeding legumes, flatulation caused by sugars and oligosaccharides can be a problem. Protein concentrates are higher in total sugar and oligosaccharides than protein isolates. Therefore, a higher proportion of isolates than of concentrates can be incorporated into feed. There is also a difference with regard to the original plant the protein is derived from. Total oligosaccharides are higher in pea and faba bean protein concentrates than in soybean concentrate, with stachyose and, to a lesser extent, raffinose as the principal α-galactoside. The oligosaccharides raffinose and stachyose are the main causes of flatulence. These are especially concentrated in air-classified protein concentrates. Protein isolates prepared by wet processing are low in oligosaccharides and therefore, flatulation is no problem when these concentrates are integrated into diets.

The air classification process concentrates some ANF, such as trypsin inhibitor activity, saponins and oligosaccharides in the protein fraction of peas, which can affect voluntary feed intake and nutrient digestibility (Frikha *et al.*, 2013). On the other hand, many of the ANFs contained in the pea seeds are sensitive to heat, and consequently, moderate heating which is applied during wet processing procedures is likely to reduce the activity and negative effects of these compounds on nutrient digestibility.

TABLE 2.9.1-7 Chemical composition (g/kg) of pea (*Pisum sativum L.*) and faba bean (*Vicia faba L.*) isolates (PPI and FBI, respectively)

Ingredient		**PPI**		**FBI**	
DM	g/kg	880	1000	880	1000
ASH	g/kg	46	52	33	37
CP	g/kg	799	908	839	953
CL	g/kg	45	51	19	22
Carbohydrates					
CF	g/kg	0.9	1	0	0
STA	g/kg	11	12	5	6
Glucose	%	0.00	0.00	0.01	0.01
Sucrose	%	0.12	0.14	0.01	0.01
Raffinose	%	0.05	0.06	0.04	0.04
Stachyose	%	0.14	0.16	0.11	0.12
Verbascose	%	0.13	0.15	0.20	0.23
Total oligosaccharides	%	0.33	0.37	0.34	0.39
Total sugars	%	0.45	0.51	0.36	0.41

BATTHY AND CHRISTISON (1984); MODIFIED

The production processes for pea protein isolate (PPI), which include drying procedures, greatly reduce the TIA concentration even if the original concentration in the raw pea was high (Le Guen *et al.*, 1995). In piglets, digestibility, both of DM and of protein, are significantly increased when PPIs were used instead of raw peas, independent of their production procedure (acid precipitation or ultrafiltration). Feed intake was not affected but weight gain was significantly enhanced when raw peas were substituted for protein isolate.

2.9.1.2.2 *Minerals*

Compared to soy protein concentrates, PPC are low in Ca and high in P (table 2.9.1-8), while the total mineral content, analysed as ash, is about the same. In pea protein isolates, the ash content varies according to the production procedure. This is obviously not the case in faba bean concentrates (Sumner *et al.*, 1981).

TABLE 2.9.1-8 Mineral content (g/kg) of pea protein concentrate (PPC), processed from white-flowered peas

Ingredient	PPC	
DM	880	1000
ASH	48	54
Ca	1.0	1.2
P	7.9	9.0
K	15.8	18.0

VALENCIA *ET AL.* (2008); MODIFIED

2.9.1.2.3 *Feed value*

Protein quality depends on the quantity of essential amino acids and their bioavailability. Pea and faba bean proteins are high in essential AAs except for sulphur containing AAs, methionine and cysteine, and for tryptophan (table 2.9.2.1-5). Moreover, methionine digestibility is lower than that of the other AAs. Therefore, the protein efficiency ratio of pea and faba bean proteins is rather poor in comparison to casein, but feeding trials with methionine supplementation to a diet with pea proteins significantly increased the protein efficiency ratio (Owusu-Ansah and McCurdy, 1991). As methionine can have an undesirable flavour and thus reduce feed intake, diets can also be supplemented with a methionine hydroxyl analogue.

The protein value also depends on the production procedures. Protein efficiency values decrease from meal to concentrate to isolate (Bhatty and Christison, 1984). Moreover, the protein solubility seems to be affected by processing procedures, especially by air classification.

TABLE 2.9.1-9 CP (g/kg) and amino acid composition (gAA/16gN) of pea (P) and faba bean (FB) protein concentrate (PC) and protein isolate (PI)

Item	PPC	PPI	FBC	FBI
CP	576	640	576	640
Lysine	7.8	8.4	6.5	7.0
Methionine	0.8	0.6	0.6	0.5
Cysteine	1.5	0.6	1.4	1.7
Threonine	3.6	3.3	3.7	3.9
Tryptophan	1.2	1.0	*	*
Arginine	9.4	11.1	11.5	12.0
Histidine	3.0	3.1	2.9	3.7
Isoleucine	3.7	4.1	3.6	4.1
Leucine	7.2	9.0	7.3	9.0
Phenylalanine	4.9	5.7	4.3	5.0
Valine	4.2	4.3	4.3	4.7

* No data.
BHATTY AND CHRISTISON (1984)

2.9.1.3 Feeding recommendations

2.9.1.3.1 *Ruminants*

Because of their good emulsifying properties, pea proteins are used as an ingredient in calf milk replacers. However, feeding trials by Bell *et al.* (1974) demonstrated that the protein digestibility of pea protein is very low in calves during the first two weeks of life. PPC, included in a milk replacer at 20% of the formula, had a digestibility coefficient of approximately 50% in calves less than 14 days old, but 74 to 80% in older calves. The digestibility of dry matter declined from 84 to 78% and protein digestibility from 67 to 63% when PPC content was raised from 7.7 to 23.1% of the dry matter in milk replacers. In older calves, the digestibility coefficient of DM was at least 89% and for CP above 81% in the highest PPC group, and 87% in the groups with less PPC inclusion (Bell *et al.*, 1974; table 2.9.1-10). In older calves, pea proteins are a suitable ingredient for milk replacers.

The low digestibility may be due to sugars and oligosaccharides in the PPC which are removed during the production of PPI. *In vitro* hydrolysis of skim milk powder, PPC and PPI with pepsin and rennin clearly demonstrated the higher digestibility of milk protein, especially at pH 4 which is the usual pH in the abomasum of young claves. Digestibility of PPC was even lower than that of PPI (Owusu-Ansah and McCurdy, 1991). Heat, generated during preparation, apparently denatures pea proteins (Bhatty, 1982) and reduces their digestibility. Moreover, both PPC and PPI contain lower levels of methionine, tryptophan, valine, and proline than milk protein.

TABLE 2.9.1-10 Effects of different levels of pea protein concentrate (PPC) in milk replacers on the digestibility of dry matter (DM) and crude protein (CP)

Calf age	Day	7–10			14–17			21–24		
Level of PPI	%	7.7	15.4	23.1	7.7	15.4	23.1	7.7	15.4	23.1
Digestibility coefficient DM	%	84	84	78	93	92	84	92	93	91
Digestibility coefficient CP	%	67	74	63	87	87	81	87	88	85

BELL *ET AL.* (1974)

Another point that restricts the use of pea proteins in calf diets is a bitter flavour that reduces feed intake in young animals.

2.9.1.3.2 *Pigs*

High protein ingredients are especially interesting in diets for piglets because of their high protein demand for muscle development. Therefore, several feeding trials with PPC and PPI have been carried out.

Ochetim *et al.*, (1980) observed poor feed utilization in piglets from 7 to 14 kg that were fed a diet containing approximately 19% of PPC as the main protein component. The protein components of the control diet were SBM and milk protein. Even though the PPC diet was supplemented with 0.3% methionine and feed intake was unaffected, daily weight gain and feed conversion ratio were significantly lower in the PPC group. Nutrient imbalance could have been a reason for the poor performance.

Le Guen *et al.* (1995) included up to 18.5% of PPI in the diet for piglets 5 weeks of age and 9 kg body weight. Compared to raw peas, daily weight gain was significantly increased in the PPI group, but it was still lower than in the groups fed casein and fish meal (290 g and 323 g weight gain, respectively). Feed intake was the same whether the diet contained casein-fish meal, raw peas or pea isolate from a cultivar with low or high trypsin inhibitor levels. The reason for the lower weight gain is most likely the lower digestibility of pea protein and the lack of essential amino acids.

Weber (2023) included smaller levels of PPC in piglet diets. Parts of high protein soybean meal were replaced by 3% PPC until three weeks, and 6% from three to six weeks after weening. No difference in piglet performance was observed between the PPC and the control group (table 2.9.1-11).

However, there was an effect on the intestinal microbiota: the proportion of streptococci was significantly reduced in the PPC groups. This could indicate a positive effect of pea protein concentrate on gut health.

TABLE 2.9.1-11 Effects of pea protein concentrate (PPC) on piglet growth and feed efficiency in two growing periods (I: 3–21 and II: 21–41 days after weening)

Growing period		Control		PPC	
		I	II	I	II
Protein ingredients in feed	%	20.5/0[1]	15/3	15/0	4/6
Body weight at start of growing period	kg	9.3	30.6	9.2	30.7
Daily weight gain	g	350	665	355	666
Feed intake per animal per day	g	451	1192	476	1217
Feed intake per kg body weight gain	kg	1.3	1.79	1.33	1.82

1 Soybean meal/pea protein concentrate.
WEBER (2023)

An inclusion rate of 3% PPC after weaning and 6% after three weeks of weening can be recommended.

2.9.1.3.3 *Poultry*

In a three-week feeding trial carried out by Velayudhan *et al.* (2019), day-old chicks were assigned to a corn-soybean meal-based basal diet with 0, 5, 10, or 15% of PPI.

The inclusion of 5 and 10% PPI did slightly, but not significantly, affect feed intake, average daily gain or feed conversion ratio. However, all three parameters were negatively affected at an inclusion rate of 15%. The weight of the small and of the large intestine were significantly higher in the PPI groups, even at an inclusion rate of 10%. It was concluded that formulating diets with increasing PPI levels decreased the growth performance of broilers. On the other hand, in this diet PPI had very small particles. That may have affected the feed intake and caused the reduced performance with higher PPI inclusion rates.

Kirn *et al.* (2023) carried out a feeding trial with PPC inclusion levels of 6.5 and 10%, replacing high-protein SBM in the broiler diet. No significant differences were found in feed efficiency, dressing percentage and carcass quality. Water intake was significantly higher for both PPC groups, but this had no effect on faeces quality, litter wetness or feet integrity at the end of the 35-day growing period. It was concluded that 10% of pea protein concentrate can be included in broiler diets.

Frikha *et al.* (2013) studied the effects of unheated and autoclaved finely ground (micronized) PPC of different pea cultivars in comparison to fermented SBM on the ileal digestibility of protein and amino acids in broilers, aged 18 to 21 days. Feed intake was highest in diets with autoclaved PPC compared to those with fermented SMB. It was lowest in groups with unheated pea protein. This was probably due to a bitter taste of raw PPC.

TABLE 2.9.1-12 Effects of rising levels of pea protein isolate (PPI) on broiler performance

Parameter		Feeding group			
		Control	5%	10%	15%
HP[1] soy in feed	%	41.5	30.9	20.3	9.8
PPI in feed	%	0	5	10	15
Overall fattening performance					
Initial body weight	g	46.0	45.6	46.2	45.2
Body weight day 21	g	704.2[a]	700.2[a]	659.7[a]	607.5[b]
Body weight gain/bird	g	658.2[a]	654.6[a]	613.6[a]	562.4[b]
Overall feed intake	g	843.3[a]	847.3[a]	805.8[a]	749.2[b]
Feed conversion ratio	kg/kg	1.28[b]	1.30[a,b]	1.32[a,b]	1.33[a]

1 High protein.
a,b $p < 0.05$.
VELAYUDHAN *ET AL.* (2019)

TABLE 2.9.1-13 Effects of pea protein concentrate (PPC) on broiler fattening

Parameter		PPC %	Growing period (days)			
			1–7	7–14	14–28	14–34
Feed intake	g/bird	0		578	1.817	2.928
		6.5		597	1.901	3.056
		10		597	1.820	2.977
Water intake	kg/bird	0		0.78	2.04[a]	2.85
		6.5		0.71	1.75[b]	2.67
		10		0.66	1.67[b]	2.51
Daily weight gain	g/bird	0	20.7	55.3	94.8	100
		6.5	22.8	56.5	97.4	116
		10	22.0	53.5	97.2	119
Feed conversion ratio	kg feed / kg weight gain	0		1.09	1.37	1.53
		6.5		1.08	1.40	1.49
		10		1.10	1.34	1.44

a,b $p < 0.05$.
KIRN *ET AL.* (2023)

Autoclaving reduced trypsin inhibitor activity (TIA) and the content of stachyose but increased that of sucrose, raffinose, and verbascose. There was little effect on the saponin content of the pea protein. Digestibility was similar for the unheated pea protein and the fermented SBM. However, autoclaving pea protein improved its overall digestibility (87.6 vs. 82.2%) and that of most AAs (e.g., 92.1 vs. 88.8% for Lys and 83.6 vs. 76.5% for Thr). The effect of the heating procedure depended on the pea cultivar on the one hand and on the remaining TIA activity. On the other hand, the fine grinding of PPC also had a positive effect on the intestinal digestibility.

2.9.2 *Hulls*

Gerhard Bellof

Grain legumes – especially grain peas, soya beans and field beans – can be processed into protein concentrates for use primarily in food production (section 2.9.1). This process produces hulls that can be used in particular in livestock feed. Using the example of grain peas, the special features of these products with regard to the intestinal health of monogastric livestock are shown below. Table 2.9.2-1 shows important ingredients for grain peas and the products made from them.

TABLE 2.9.2-1 Ingredients and energy content of grain peas and the pea products made from them – pea hulls and pea protein concentrate – (data based on 88% DM)

Ingredient		Peas	Pea hulls	Pea protein concentrate
Dry matter	g	880	880	880
Crude ash	g	26	25	44
Crude protein	g	207	46	724
Lysine	%	1.54	0.28	5.35
Methionine	%	0.18	0.05	0.70
Cysteine	%	0.31	0.08	0.74
Threonine	%	0.74	0.16	2.68
Tryptophan	%	0.17	0.04	0.73
Valine	%	0.99	0.20	3.51
Isoleucine	%	0.87	0.17	3.53
Leucine	%	1.43	0.27	5.91
Arginine	%	1.58	0.21	5.93
Crude fibre	g	58	451	14
$aNDF_{om}$	g	95	635	19
ADF_{om}	g	78	558	17
ADL	g	11	9	11

TABLE 2.9.2-1 Ingredients and energy content of grain peas (*cont.*)

Ingredient		Peas	Pea hulls	Pea protein concentrate
NFC	g	532	171	13
Crude fat	g	20	4	13
Starch	g	449	85	*
Sugar	g	48	*	13
AME_n	MJ	12.0	2.3	14.1

* Below detection limit.
KIRN *ET AL.* (2023)

2.9.2.1 Definition and properties of dietary fibre

Pea hulls are coming into focus due to their dietary fibre content. The term "dietary fibre" is generally understood to cover the complex mixture of indigestible polysaccharides such as cellulose, hemicelluloses, oligosaccharides, resistant starch, pectins, lignin and related plant substances, which occur in plants mainly as cell wall material. These dietary fibres resist digestion and absorption in the small intestine and are partially or completely fermented into short-chain fatty acids primarily in the large intestine by microorganisms (Freitag, 2021; Tosh and Yada, 2010). In addition to short-chain fatty acids (SCFA), gases are also among the fermentation products. While the gases can cause digestive problems due to flatulence, the SCFA support the health of the intestinal mucosa (Tosh and Yada, 2010).

The carbohydrates ingested with food form a diverse group of compounds that differ considerably in terms of their chemical structure and physicochemical properties (Bach Knudsen *et al.*, 2018). Dietary fibres are divided into soluble and insoluble dietary fibres according to their solubility (Dusel, 2018). Lignin, cellulose and some hemicelluloses fall into the category of insoluble dietary fibres (Kumari and Deka, 2021; Tosh and Yada, 2010). Soluble fibres include resistant starch, galactomannan oligosaccharides, hemicellulose, ß-glucans, pectins and fructans (Freitag, 2021; Kumari and Deka, 2021; Tosh and Yada, 2010). These two fractions are characterised by different physiological functions and nutritional-physiological benefits (Tosh and Yada, 2010). The soluble fibres are the most easily fermented, while the insoluble fibres are fermented more slowly or not at all (Bach Knudsen *et al.*, 2018).

Starch and dietary fibre are the main components of peas. The dietary fibre in peas comes from both the seed coat (outer fibre) and the cotyledon (inner fibre). The total fibre content of peas is 14% to 26% of the dry matter, of which the insoluble fibre accounts for 10% to 15% and the soluble fibre for 2% to 9% (Dahl *et al.*, 2012).

The pods make a significant contribution to the insoluble fibre content of whole pulses. The removal of the hull during roller milling of peas reduces the insoluble fibre content in the dry matter from 11.3% in the seed to 5.3% in the resulting flour (Tosh and Yada, 2010).

The seed coat of the pea contains mainly the water-insoluble polysaccharide cellulose, while the cotyledon fibre consists of polysaccharides with varying degrees of solubility, including cellulose, hemicelluloses and pectins (Dahl *et al.*, 2012). It should be emphasised that both grain peas and pea hulls have low lignin contents (Freitag, 2021; Tosh and Yada, 2010).

Dietary fibre data reported as "crude fibre" in the form of acid detergent fibre and neutral detergent fibre is of limited value. Crude fibre analysis isolates only part of the insoluble fibre fraction, so that the total fibre content is greatly underestimated.

Nutritionally important soluble fibre components are also lost in the ADF and NDF analysis. The characterisation of cell wall polysaccharides in legumes is essential to better understand their function as dietary fibres (Tosh and Yada, 2010).

2.9.2.2 Importance of dietary fibre for monogastric animals

"Dietary fibre" is the most important dietary factor that influences the digestion and absorption processes at all sites of the gastrointestinal tract and the secretion of fluids from the pancreas and liver (Bach Knudsen *et al.*, 2018). They serve as functional ingredients and have various positive physiological effects on the health of monogastric animals (Tosh and Yada, 2010). The effect of dietary fibres on monogastric animals begins passively with feed intake, includes intestinal microbiology and its metabolites, continues with immunology and ends with animal performance (Schedle, 2018). Dietary fibre fulfils various functions in the animal body through different effects on animal health and well-being (Figure 2.9.2-1) (Freitag, 2021).

However, the increased passage rate does not necessarily apply to poultry, as the gizzard selectively retains various feed particles and the passage rate in the gizzard is reduced by insoluble fibres (Hetland *et al.*, 2004). Soluble fibres can reduce nutrient absorption and disrupt digestive processes in the small intestine due to their pronounced water retention capacity and overall swelling capacity (Bach Knudsen *et al.*, 2018; Freitag, 2017). In poultry, water-soluble non-starch polysaccharides favour the proliferation of *C. perfringens* and predispose young chicks to necrotic enteritis (Pan and Yu, 2013). Dietary fibres can also have a negative effect on the digestibility of nutrients and energy (Jha and Leterme, 2012). Lignified dietary fibres are not broken down, while insoluble fibres, especially those with a high lignin content, result in a reduction in feed conversion. The digestibility of the ration decreases by around 1% per per cent NDF (Freitag, 2021; Freitag, 2017).

Due to the general prebiotic effect of dietary fibre on the gut microbiota, dietary fibres have positive effects on the health of monogastric animals. The complex carbohydrates show positive effects on gut health, as they have an influence on the

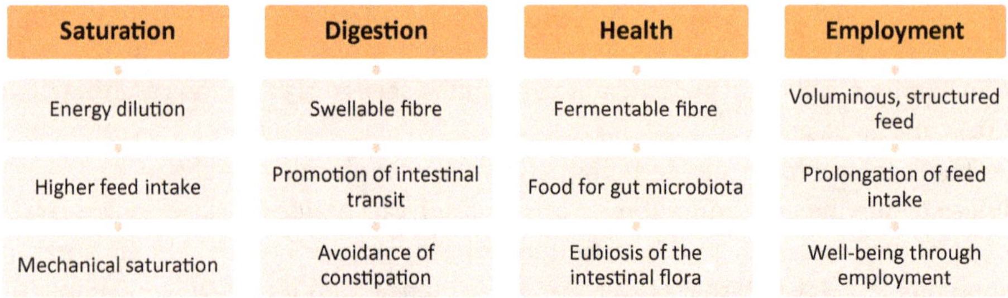

FIGURE 2.9.2-1 Effects of dietary fibre on the welfare and animal health of monogastric animals
FREITAG (2021); MODIFIED

composition of the gut microbiota, growth stimulation, metabolism and the activities of lactic acid bacteria, bifidobacteria and other microbial strains (Clemente and Olias, 2017). It is known that a beneficial microbial community modulates the host's immune system and influences the host's organ development and metabolism (Oakley *et al.*, 2014). The gut microbiome promotes the positive development of the intestinal mucus layer and influences intestinal morphology and the activity of digestive enzymes (Oakley *et al.*, 2014; Pan and Yu, 2013).

The dietary fibres fermented anaerobically in the large intestine by the intestinal microbiota release a large number of metabolites. These include short-chain fatty acids, especially acetate, propionate and butyrate, as well as essential vitamins that influence local and systemic host physiology (Clemente and Olias, 2017; Pan and Yu, 2013). The short-chain fatty acids fermented by the intestinal microbiota have an anti-inflammatory effect, promote the growth of the "probiotic" intestinal microbiota, stimulate the proliferation of epithelial cells in the colon, stimulate intestinal mucus formation, form a barrier for pathogenic germs and stabilise overall intestinal health (Che *et al.*, 2014; Freitag, 2021; Freitag, 2017). The short-chain acids produced as a result of microbial fermentation not only have an impact on intestinal health, but also serve as a source of energy for the animal and are involved in a variety of metabolic processes (Freitag, 2021; Pan and Yu, 2013).

Bacterial fermentation increases with increasing age of the animal. There is also a clear successive change in the taxonomic composition of the microbiome during growth (Oakley *et al.*, 2014; Pan and Yu, 2013).

The development of microorganisms in the gastrointestinal tract can be influenced by a targeted ration design with "dietary fibre" due to its mode of action (Freitag, 2021). The amount and type of dietary fibre determines the bacterial flora in the large intestine (Eastwood and Passmore, 1983). In vivo experiments show that oligosaccharides in particular significantly influence the growth of bifidobacteria in the large intestine. Oligosaccharides, which occur as soluble carbohydrates in significant concentrations in peas and other legumes, are therefore potential prebiotics (Dahl *et al.*, 2012; Tosh and Yada, 2010).

With the positive health effects described, dietary fibres improve key aspects of gut health (Clemente and Olias, 2017). The ability of dietary fibre to positively influence gut health is particularly noteworthy, as the gut is the largest immune organ in the living organism (Dusel, 2018). Consequently, dietary fibre appears to have protective functions against the main problems of intensive diets in monogastric animals (Schedle, 2018).

There are currently no specific recommendations for the use of dietary fibre. Specific recommendations cannot be considered standardised, as they often refer to the total crude fibre according to the Weender feed analysis. Certain proportions of non-enzymatically digestible, easily soluble fibres cannot be detected by existing analysis methods (Freitag, 2021). In order to be able to address the physico-chemical properties of dietary fibre, the analysis methods must be expanded to include the values for soluble and insoluble fibres and fermentation as well as viscosity so that certain feeding requirements can be met (Dusel, 2018).

2.9.2.3 Feeding recommendations

2.9.2.3.1 *Pigs*

Weber *et al.* (2021) tested the use of pea hulls in the feeding of fattening pigs. The effects of pea hulls on the performance characteristics of fattening pigs were investigated in a three-phase feeding trial. Hulls from white-flowering varieties were used. Pea hulls were added to the complete feed mixtures for the three test groups. In the pre-fattening phase, the proportion of pea hulls was 2.5% in all three groups. In the final fattening stage, the addition levels were 2.5%, 5% or 70%. A batch mixer was used to prepare the mixtures for the middle fattening period. The feed mixtures of the control group contained wheat bran and wheat semolina bran as a source of fibre. No significant differences were found between the test groups and in the individual fattening periods for the characteristics of daily weight gain and feed consumption. In contrast, the feed consumption per kg gain (total fattening) was significantly improved by the use of 2.5% pea hulls during fattening.

In a feeding trial with weaned piglets, Weber (2023) investigated the use of peas, pea hulls and pea protein concentrate. A total of 348 piglets (origin: Pietrain × Topigs) were used. The piglets were on average 27 days old at the start of the trial. Two complete feed mixtures were used in each of the 6 feeding groups (piglet rearing feed (FAF) 1 and 2). The experimental design can be seen in Table 2.9.2-2 below. In addition to the peas and pea protein concentrate listed there, soya extraction meal (SES-HP) served as the main source of protein. SES-HP was the sole protein feed for the control group. Barley, wheat and rapeseed oil were used as the main energy sources in the feed mixtures. The biological performance of all animals and the microbiome of selected animals were analysed.

In this trial, the use of the various pea products in the mix proportions used had no influence on the biological performance of the piglets. This means that even

TABLE 2.9.2-2 Use of peas and pea products in the feeding trial with weaned piglets (figures in % in complete feed)

Feeding group		Piglet rearing feed 1			Piglet rearing feed 2		
No.	Designation	P	PPC	PH	P	PPC	PH
1	C	–	–	–	–	–	–
2	P 20	10	–	–	20	–	–
3	PPC 3/6	–	3	–	–	6	–
4	PH 3/6	–	–	3	–	–	6
5	P 10/20 + PH 3/6	10	–	3	20	–	6
6	PPC 3/6 + PH 3/6	–	3	3	–	6	6

P: peas; PPC: pea protein concentrate; PH: pea hulls; C: control
WEBER (2023)

doubling the crude fibre content in FAF 2 did not reduce performance. The reason for this could be the increased fermentation due to a higher activity and abundance of carbohydrate-fermenting bacteria in these feeding groups. The use of pea hulls led to a significant increase in the genus *Faecalibacterium*, while *Streptococcus* showed significantly reduced abundances in the same experimental groups and when pea concentrate was used. On the other hand, *Lachnospira* was a subdominant genus that became the dominant genus in the mixed diets. In comparison to other feeding groups, numerical differences were particularly noticeable in the experimental group with peas. Here, an increase in *Streptococcus* and *Prevotella_7* was observed with a reduction in *Lactobacillus*.

When analysing the relationships between the microbiota and fermentable feed components, it is noticeable that the grain content showed more significant correlations than the pea products. It was found that pea hulls in particular showed positive relationships with bacteria that typically ferment carbohydrates in the large intestine (*Prevotella, Lachnospiraceae, Ruminococcus*), while *Streptococcus* led to significant negative correlations with pea hulls and pea protein concentrate.

As a fibre carrier, the hulls of grain legumes have advantages due to their lower phosphorus content compared to rapeseed meal and wheat bran (Weber 2023).

2.9.2.3.2 *Poultry*

Kirn *et al.* (2024) investigated the use of peas, pea hulls and pea protein concentrate in complete feed mixtures for broiler fattening. A total of 800 male broilers of the Ross 308 genotype (breeding company Aviagen) were housed. The day-old chicks were divided into 10 experimental groups with 4 replicates of 20 birds each. The trial was divided into a starter phase P1 (day of life 1–14) and a fattening phase P2

TABLE 2.9.2-3 Design of the study and mixture proportions (in %) of peas (P), pea protein concentrate (PPC) and pea hulls (PH) in the complete feed mixtures for broilers

Feeding group		Starter phase (P1)			Finishing phase (P2)		
No.	Designation	P	PPC	PH	P	PPC	PH
1	Control	–	–	–	–	–	–
2	P 20	20	–	–	20	–	–
3	P 30	30	–	–	30	–	–
4	PPC 6.5	–	6.5	–	–	6.5	–
5	PPC 10	–	10	–	–	10	–
6	PH 3/4.5	–	–	3	–	–	4.5
7	PH 4.5/6	–	–	4.5	–	–	6
8	P 20 + PH 3/4.5	20	–	3	20	–	4.5
9	PPC 10 + PH 3/4.5	–	10	3	–	10	4.5
10	PPC 10 + PH 4.5/6	–	10	4.5	–	–	6

KIRN *ET AL.* (2023)

(day of life 14–34). The experimental design is shown in Table 2.9.2-3. In addition to the peas and pea protein concentrate listed there, soya extraction meal (SES-HP) served as the main source of protein. SES-HP was the sole protein feed for the control group.

Grain maize, wheat and rapeseed oil were used as the main energy sources in the feed mixtures. The relevant characteristics of fattening and slaughter performance as well as the microbiome of slaughter animals were analysed.

The authors draw the following conclusions from the results of the feeding trial:

1) The targeted use of peas and pea protein concentrate leads to a significant reduction in the proportion of soya extraction meal in the feed mixtures without causing any loss of performance. This means that the use of peas can reduce the critical import of soya products from overseas. In addition, the high starch content of peas reduces the proportion of energy-supplying feed in the mix.
2) The use of pea hulls causes increasing crude fibre content and cell wall components in complete feed mixtures. Consequently, pea hulls are suitable as a fibre-dense feed for targeted balancing in order to avoid adverse effects on the microbiome.
3) The use of peas in mixtures of up to 30% has no negative effects on the fattening and slaughter performance of broilers. Nevertheless, the use of 30% peas in the ration tends to reduce the biological performance of broilers. This may be due to the poorer digestibility of pea starch.

4) Towards the end of fattening, the addition of pea hulls alone and the combination of peas (20%) and pea hulls in an amount of up to 6% leads to higher feed consumption, while the use of pea hulls alone also results in lower feed consumption. It is possible that the high concentrations of crude fibre and cell wall components in the complete feeds have negative effects on nutrient digestibility. The broilers could potentially compensate for the poorer nutrient digestibility by consuming more feed.
5) The use of pea protein concentrate leads to an overall increase in broiler performance. Primarily in the combination of pea protein concentrate (10%) with pea hulls (up to 6% mixture proportion), the broilers tended to achieve higher live weights and daily weight gains and improved feed consumption. The results indicate a high amino acid digestibility of the pea protein concentrate. In addition, the structural components may possibly lead to an improvement in starch digestibility. The effect of improved nutrient digestibility appears to be enhanced by moderate proportions of pea hulls.
6) The broilers of the variant with peas (20%) and pea hulls (up to 4.5%) show the highest water consumption overall. This is due to the comparatively high potassium content in the complete feed mixtures. The use of pea protein concentrate, on the other hand, leads to low potassium levels in the rations and significantly reduces water consumption.
7) Peas and pea products have no negative influence on the relevant characteristics of slaughter performance and the weights of liver, gizzard and abdominal fatness.
8) Peas and pea products have no influence on the levels of short-chain fatty acids in the digesta of the ileum and caecum. The use of peas and pea products tends to lead to increased levels of propionic acid, n-butyric acid and total fatty acids in the digesta of the caecum. Peas and pea products potentially have a positive influence on the metabolism of bacteria.
9) Peas and pea products have no effect on the relative abundance of bacterial 16S rDNA in the digesta of the ileum and on the richness, evenness and Shannon index of the gut microbiome in the ileum.
10) Statistically confirmed differences in the relative abundance of bacterial 16S rDNA in the digesta of the caecum were found for certain genera. In addition, the use of peas in combination with pea hulls (E20 + ES4.5) leads to a high richness, evenness and a high Shannon index of the microbiome in the caecum. It is possible that the non-starch polysaccharides from peas and pea hulls lead to an increased diversity of the gut microbiome.
11) In the ileum, the use of peas and pea products tends to reduce the relative abundance of the potentially harmful bacteria *Erysipelatoclostridium, Escherichia/Shigella* and *Helicobacter*. The reduced relative abundance of *Erysipelatoclostridium* can be continued in the caecum.

Presumably, the use of peas and pea products leads to a stimulation of the commensal intestinal microbiota, particularly in the ileum. Peas and pea products could therefore potentially improve the intestinal health of broiler chickens.

References

ADM, 2024: Ingredients catalog, feed and pet food. online 30.04.2024. https://www.adm.com/globalassets/products--services/animal-nutrition/products/macro-ingredients/adm-feed-ingredients-catalog.pdf.

Bach Knudsen K.E., Hedemann M.S., Laerke H.N., 2018: Using animal models to understand the mode of actions of carbohydrates and phytochemicals during digestion and absorption. In: Kohlenhydrate in der Tierernährung, Tagungsband 17. BOKU-Symposium Tierernährung, Universität für Bodenkultur Wien, ISBN 978-3-900932-52-7.

Bell J.M., Royan G.F., Youngs C.G., 1974: Digestibility of pea protein concentrate and enzyme-treated pea flour in milk replacers for calves. Canadian j. Anim. Sci., 54(3): 355–362.

Bhatty R.S., 1982: *In Vitro* Hydrolysis of skim milk and pea proteins by pepsin and rennin. Can. Institute food sci. Techn. J. 15 (2): 101–108.

Bhatty R.S., Christison G.I., 1984: Composition and nutritional quality of pea (*Pisum sativum L.*), faba bean (*Vicia faba L. spp. minor*) and lentil (*Lens culinaris Medik*) meals, protein concentrates and isolates. Plant foods for human nutrition, 34: 41–51.

Boukid F., Rosell C.M., Castellari M., 2021: Pea protein ingredients: a mainstream ingredient to (re) formulate innovative foods and beverages. Trends in food science & technology, 110: 729–742.

Clemente A., Olias R., 2017: Beneficial effects of legumes in gut health. Current opinion in food science 2017, 14, 32–36, DOI: 10.1016/j.cofs.2017.01.005.

Dahl W., Foster L, Tyler R., 2012: Review of the health benefits of peas (Pisum sativum L.). British journal of nutrition 2012 Aug, 108, 3–10. DOI: 10.1017/S0007114512000852.

Dazert D., 2004: Untersuchung der Entölbarkeit von Sojamehl zur Herstellung von Sojaproteinprodukten. Technische Universität Berlin, Thesis.

Dusel G., 2018: Die Nutzung der Faser in der Ernährung der Sau. In: Kohlenhydrate in der Tierernährung, Tagungsband 17. BOKU-Symposium Tierernährung, Universität für Bodenkultur Wien, ISBN 978-3-900932-52-7.

Eastwood M.A., Passmore R., 1983: Dietary fibre. The lancet 1983, 2(8343), 202–206, DOI: 10.1016/s0140-6736(83)90183-6.

Freitag M., 2017: Rohfaser, Ballaststoffe und Co. Jahresbericht 2017, Hrsg.: Erzeugerring Westfalen eG, Senden-Bösensell, https://www.fh-swf.de/media/neu_np/fb_aw_2/dozentinnen/professorinnen_2/freitag/publikationen_pdf/neu_1/2018_Ballaststoffe_und_Co_ERW_2017.pdf Stand: 22:04.2022.

Freitag M., 2021: Bedeutung von Fasern für die Tierernährung. Editors: Durst J., Freitag M., Bellof G., 2021: Futtermittel für landwirtschaftliche Nutztiere. DLG- Verlag, ISBN 978-3-7690-0852-4, Frankfurt am Main, 1. Auflage, 482–500.

Frikha M., Valencia D.G., De Coca-Sinova A., Lazaro R., Mateos G.G., 2013: Ileal digestibility of amino acids of unheated and autoclaved pea protein concentrate in broilers. Poultry science, 92(7): 1848–1857. doi:10.3382/ps.2013-03007.

Hetland H., Choct M., Svihus B., 2004: Role of insoluble non-starch polysaccharides in poultry nutrition. World's poultry science journal 2004, 60(4), 415–422, DOI: 10.1079/WPS200325.

Jha R., Leterme P., 2012: Feed ingredients differing in fermentable fibre and indigestible protein content affect fermentation metabolites and faecal nitrogen excretion in growing pigs. Animal 6 (4), 603–611, DOI: 10.1017/S1751731111001844.

Kirn A.I., Weindl P.A., Bellof G., Vahjen M., Zentek J., 2023: Erbsen und Erbsenprodukte in der Hühnermast unter besonderer Berücksichtigung der Darmgesundheit. Abschlussbericht, UFOP-Schriften 2023. https://www.ufop.de/medien/downloads/agrar-info/forschung.

Kirn A.I., Vahjen M., Weindl P.A., Hofmann P., Zentek J., Bellof G., 2024: Effect of peas and pea products in diets for broiler chickens with consideration of the intestinal microbiota. Journal of applied poultry research. doi/10.1016/j.japr.2023.100401.

Kumari T., Deka S., 2021: Potential health benefits of garden pea seeds and pods, British journal of nutrition 2012, 108 (1), 3–10, DOI:10.1017/S0007114512000852.

Le Guen M.P., Huisman J., Guéguen J., Beelen G., Verstegen M.W.A., 1995: Effects of a concentrate of pea antinutritional factors on pea protein digestibility in piglets. Livestock production science 44(2): 157–167. doi:10.1016/0301-6226(95)00053-4.

Oakley B., Lillehoj H., Kogut M., Kim W.K., Maurer J., Pedroso A., Lee M., Colett S., Johnson T., Cox N., 2014: The chicken gastrointestinal microbiome. FEMS microbiology letters, 360(2), 100–112, DOI: 10.1111/1574-6968.12608.

Ochetim S., Bell J.M., Doige C.E., 1980: The feeding value of tower rapeseed for early weaned pigs. II. Effects of supplementation of cooked seed with iodinated casein and pea protein concentrate. Can. J. Anim. Sci. 60: 423–432.

Owusu-Ansah Y.J., McCurdy S.M., 1991: Pea proteins: a review of chemistry, technology of production, and utilization. Food reviews international, 7(1): 103–134.

Pan D., Yu Z, 2013: Intestinal microbiome of poultry and its interaction with host and diet. Gut microbes 2014, 5(1), 108–19, DOI:10.4161/gmic.26945.

Rao A., Shallo H.E., Ericson A.P., Thomas R.L., 2002: Characterization of soy protein concentrate produced by membrane ultrafiltration. J. Food sci. 67(4): 1412–1418.

Schedle K., 2018: Dietary fiber: more than a poor nutrient in swine nutrition. In: Kohlenhydrate in der Tierernährung, Tagungsband 17. BOKU-Symposium Tierernährung, Universität für Bodenkultur Wien, ISBN 978-3-900932-52-7.

Sumner A.K., Nielsen M.A., Youngs C.G., 1981: Production and evaluation of pea protein isolate. Journal of food science, 46(2): 364–366.

Stein H.H., Berger L.L., Drackley J.K., Fahey G.C. Jr., Hernot D.C., Parsons C.M., 2008: Soybeans | nutritional properties and feeding values of soybeans and their coproducts. soybeans / chemistry, production, processing, and utilization. Editors: Johnson L.A., White P.J., Galloway E., AOCS Press, 613–660.

Tosh S.M., Yada S., 2010: Dietary fibres in pulse seeds and fractions: Characterization, functional attributes, and applications. Food research international, 43, 450–460, DOI: 10.1016/j.foodres.2009.09.005.

Valencia D.G., Serrano M.P., Centeno C., Lázaro R., Mateos G.G., 2008: Pea protein as a substitute of soya bean protein in diets for young pigs: Effects on productivity and digestive traits. 118(1–2), 0–10. doi:10.1016/j.livsci.2008.01.018.

Velayudhan D.E., Mejicanos G.A., Nyachoti C.M., 2019: Evaluation of pea protein isolates as a protein source for broilers. Poultry science 98: 803–810.

Weber M., Mäurer H., von Klopoteck E., Bönisch K., 2021: Einsatz von Erbsenschalen in der Schweinemast. Abschlussbericht, UFOP-Schriften 2021. https://www.ufop.de/medien/downloads/agrar-info/forschung.

Weber M., 2023: Einfluss verschiedener Erbsenprodukte in der Ferkelfütterung. Abschlussbericht, UFOP-Schriften 2023. https://www.ufop.de/medien/downloads/agrar-info/forschung.

CHAPTER 3

Small-grain legumes

3.1 Importance and cultivation of small-grain legumes

Stephan Hartmann

3.1.1 *Importance*

In many regions, intensive forage production is the basis for land-based animal husbandry, alongside appropriate grassland management. Forage crops include clover-like fodder plants, lucerne, forage grasses and their mixtures. In a broader sense, this also includes silage maize and fodder beet.

The topics of this section are red clover (*Trifolium pratense* L.), lucerne (*Medicago* L. *partim*), white clover (*Trifolium repens* L.), Alsike clover (*Trifolium hybridum* L.), Berseem clover (*Trifolium alexandrinum* L.), Persian clover (*Trifolium resupinatum* L.), bird's-foot trefoil (*Lotus corniculatus* L.), *crimson clover* (*Trifolium incarnatum* L.) and sainfoin (*Onobrychis viciifolia* Scop.). Of these, only red clover, lucerne, Berseem clover and Persian clover are grown in pure seed. As a rule, however, these and all other small-grain legumes are grown in mixtures with grasses and/or other small-grain legumes. Cultivation as a mixture offers greater yield security and higher yields are possible, especially in combination with grasses. The reasons for this are the spreading of risk in mixtures of species by extending the ecological breadth and the higher yield potential of grasses under suitable conditions. (Sleugh *et al.*, 2000; Foster 2024, Suter *et al.*, 2015; Elgersma and Søegaard, 2016).

This is why small-grain legumes and their mixtures with grasses still play an important role in forage production today and are essential for organic farming. Their growing share has positive effects on the area development of forage legumes. In conventional farming, forage crops are always in competition with the high-yielding but low-protein energy source silage maize, which has benefited most from the steep rise in demand for biomass for biogas plants. The production volume, cultivation importance and value of clover, clover-grass and lucerne vary greatly from region to region in Germany. The proportion of clover or lucerne and clover or lucerne-grass mixtures in field forage production is increasing from north to south. The reason for this is the increasingly continental climate. The proportion of legumes in the mixture mitigates the increasing risk of phases of temporary drought stress along this gradient and its negative consequences on the yield and quality of the grasses, for example through stress-induced ear emergence.

 | DOI:10.3920/9789004695047_004

The advantages of small-grain legumes in crop rotation are:
- the excellent promotion of soil life through root and stubble residues,
- the opening up of the subsoil through the deep root network,
- shading that promotes fermentation for large parts of the year,
- nitrogen enrichment,
- Crop rotation loosening in the case of a strongly cereal-orientated cultivation ratio,
- fertility-enhancing enrichment with organic matter during recultivation measures.
- Contribution to the regulation of undesirable species, as these are generally not re-cut and are also subject to competition from the clover-grass stand in the longer term.

The favourable effect of legumes on the soil and subsequent crop is explained in detail in section 2.1. Among the clover species, red clover is the most important on the field. Persian and Berseem clover are mainly used as catch crops. In dry areas, lucerne is the predominantly cultivated legume species.

The small-grain legumes can be used in a variety of ways: grazing, green feeding, silage, haymaking or hot-air drying.

While lucerne is the most important green legume in fodder production worldwide and in Europe, red clover occupies this position in Germany (table 3.1-1).

Small-grain legumes and their mixtures with grasses with a legume content of more than 80% accounted for approx. 3.0% of arable land in Germany in 2023 (DeStatis, 2023).

However, the distribution of cultivation varies greatly within the EU (see table 3.1-1) and also in Germany. For example, three federal states dominate cultivation with more than 50% of the national cultivation area.

FIGURE 3.1.1-1
Seed of red clover as an example for "small-grain legumes"
PHOTO: © A. WOSNITZA, LFL BAYERN

TABLE 3.1.1-1 Certified seed sales [t] of the most important small-grain legumes for forage, grassland and catch crops in Europe in 2021

Country	Lucerne	Red clover	White clover	Berseem clover
Bulgaria	81.00			84.00
Begium	–	0.01	–	–
Czech	66.70	828.90	–	–
Denmark	72.30	177.90	2,908.80	0.70
Germany	479.30	840.40	162.30	165.10
Spain	1,285.20	–	–	–
Estonia	5.00	107.50	1.60	–
France	2,733.80	968.00	–	135.60
Greece	273.70	–	–	10.50
Switzerland	–	59.30	–	–
Croatia	96.70	2.70	1.90	–
Iceland	–	–	–	–
Ireland	–	–	–	–
Italy	14,413.70	581.50	261.10	6,503.30
Latvia	1.60	302.00	237.00	–
Lithunia	0.30	970.30	500.20	–
Luxembourg		1.20	–	–
Hungary	946.70	22.10	–	12.10
Malte	–	–	–	–
Netherlands	–	–	–	–
Norway	–	185.90	6.80	–
Austria	5.60	54.50	–	–
Poland	–	–	–	–
Portugal	–	–	–	107.70
Romania	2,787.00	57.50	–	–
Slovenia	–	0.90	–	–
Slovakia	–	29.30	–	–
Serbia	–	–	–	–
Finland	–	11.00	–	–
Sweden	–	–	–	–
Cyprus	–	–	–	–
Summe	23,248.60	5,200.91	4,079.70	7,019.00

DATA FROM EUROPEAN SEED CERTIFICATION AGENCIES ASSOCIATION, 2024; HTTPS://ESCAA.ORG/INDEX/ACTION/PAGE/ID/9/TITLE/CERTIFIED-SEED-QUANTITIES

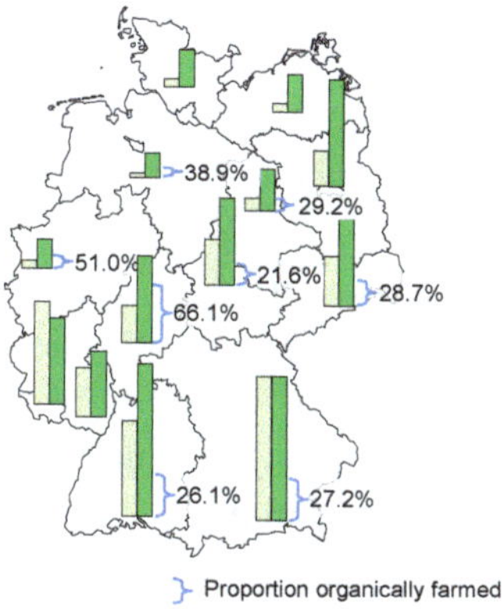

Federal states of Germany	'03 %	'17 %
Bavaria	4,7	4,7
Saarland	3,3	2,8
Baden - Württemberg	3,0	4,9
Saxony	1,6	2,8
Rhineland - Palatinate	1,6	2,1
Thuringia	1,5	2,8
Hesse	1,2	2,9
Brandenburg	1,1	3,4
Saxony - Anhalt	0,4	1,3
North Rhine - Westphalia	0,3	0,9
Meck - Western Pomerania	0,3	1,2
Schleswig - Holstein	0,3	1,2
Lower Saxony	0,1	0,8
Germany	1,5	2,3

LfL

Hartmann - IPZ 4b – 2024

FIGURE 3.1.1-2 Share of clover, lucerne and their mixtures with grasses in arable land in Germany increases with continental influence (2003 and 2017)

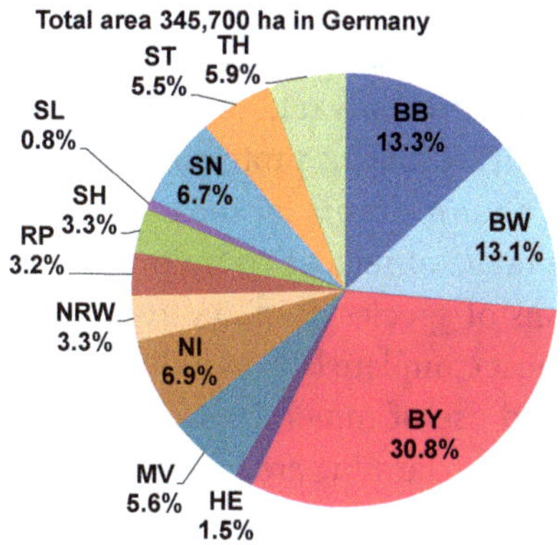

Abbreviation	Federal states of Germany
BY	Bavaria
SL	Saarland
BW	Baden - Württemberg
SN	Saxony
RP	Rhineland - Palatinate
TH	Thuringia
HE	Hesse
BB	Brandenburg
ST	Saxony - Anhalt
NRW	North Rhine - Westphalia
MV	Meck - Western Pomerania
SH	Schleswig - Holstein
NI	Lower Saxony

Source: DeStatis 6 February 2023 Wachstum und Ernte - Feldfrüchte - Fachserie 3 Reihe 3.2.1 - 16/2022 (latest edition, discontinued on a report-by-report basis)

LfL

Hartmann - IPZ 4b – 2024

FIGURE 3.1.1-3 Share of the federal states in the cultivation of small-grain legumes (e.g. clover, lucerne) and their mixtures from 80% legumes in 2022

In Germany, starting with Bavaria's protein initiative in 2011 (Lindermayer, 2014; Bayerisches Staatsministerium für Ernährung, Landwirtschaft und Forsten, 2015), protein initiatives were also launched in other federal states and the federal government in 2012 and later (BMEL, 2012). All pursue the goal of reducing dependence on imported protein animal feed. One lever here is the improvement and return to coarse fodder with higher crude protein content (grassland and field fodder cultivation). On the one hand, this is due to the high extent of cultivation (around one third of Germany's agricultural land is grassland and 3% of arable land is legume-oriented field fodder cultivation) and on the other hand to the high crude protein yields per hectare, since, in contrast to large-grain legumes, the crude protein of the whole plant is utilised and not just a comparatively small proportion (grains). One aim is to raise awareness of the benefits of forage crops and to expand their cultivation. The success can be seen in figures 3.1-1 and 3.1-2. The protein crop strategy is aimed equally at conventional and organic cultivation (BLE, 2024). Other EU member states are also taking measures to promote domestic protein crop production (see Chapter 2), implemented as part of Common Agricultural Policy strategy plans from 2023. Within this framework, the EU is also promoting the breeding of legumes, including small-grain legumes (Boosting innovation in breeding for the next generation of legume crops for Europe). In this context, a knowledge base was also created on the Internet (Legumhub, 2024). The acreage of forage crops depends mainly on utilisation by ruminants. Due to the significantly less favourable C:N ratio – for the biogas plant – this crop rotation element benefits significantly less from the option of utilisation in biogas plants. In order to be able to expand cultivation even with declining livestock numbers, it is therefore essential to expand utilisation. One way is to valorise and differentiate through the technical separation of leaves and stems or the pressing of the plant sap. The leaves and pressed juices obtained and preserved in this way achieve a high protein density and quality and can be used as protein feed in concentrate mixtures for both high-yielding ruminants and monogastric livestock (Bellof *et al.*, 2021). Details can be found in the following sections (3.2, 3.4 and 3.5). This is particularly relevant for organic farming, as it can help to close the gap in high-quality, domestic protein feed.

The right mix of species with lucerne and/or clovers for the region and the specific production goal varies over a wide range of recipes (from pure cultivation to 8–10 species) in terms of the proportions of species used and the recommended varieties of these species. Due to the almost complete lack of indications for chemical plant protection and the very limited use of mineral nitrogen fertilisers, the differences between conventional and organic farming are negligible.

For this reason, the types and their use are described in general terms below.

3.1.2 *Cultivation*

3.1.2.1 Lucerne (*Medicago* L. *partim*, *Medicago sativa* L., *Medicago* × *varia*)

It has a typical taproot. The main roots penetrate quickly and deeply into the soil, over 4 metres deep if there is sufficient permeability. The strongest branching is achieved in the soil layer between 30 and 50 cm. The lateral and depth extension of the root network depends on the soil and water conditions. In dry locations, a greater depth but less lateral expansion is observed. In moist locations, a shallower but more widely branched root system develops. In lucerne, the root mass is also the reservoir in which assimilates are collected from the beginning of flowering for rapid resprouting. At the root head of the lucerne, the shoots for the new growth after cutting develop from above-ground renewal buds. This above-ground bud location explains the sensitivity of lucerne to deep biting and cutting. At the beginning of winter, the shoots freeze back as they are not “pulled in” as with red clover. While the blue-violet flowering seed lucerne dominates cultivation in warmer EU countries such as Italy or France, all varieties registered in the German variety list are so-called bastard lucerne. They are true species bastards of the high-yielding lucerne also named Alfalfa (*Medicago sativa* L.) with the hardy yellow-flowering sickle lucerne (*Medicago falcata* L.) and therefore flower colourfully – the colours of the individual flowers in a variety range from blue to variegated to yellow in these population varieties. This is also recognisable in the habitus of coarse-stemmed, tall-growing, less hardy variety types and, on the other hand, fine-stemmed, less sturdy, hardy variety types. The latter offer advantages for wilting and haymaking due to faster drying. The main advantages of lucerne are its drought tolerance, high crude protein content, high acceptance by ruminating livestock and its persistence, which is better than that of red clover.

In pure cultivation, the utilisation period ranges from the year of sowing and one to three main years of use. In good (low-grass) locations, more than five years of utilisation are also possible. Areas with this length of utilisation do not lose their arable status. In addition to typical lucerne grass mixtures, which are easier to ensile due to the sugar content of the grass component, admixture with clover-grass mixtures in "non-typical" lucerne locations can also be useful. In such mixtures, the lucerne increases the yield, especially in dry summer months or on the dry parts of a field.

"Lucerne loves a warm head and does not tolerate wet feet!" This old rule characterises the main requirement of lucerne: a warm, sunny, predominantly dry climate and under no circumstances constant high humidity. However, the heat requirements are not so high that its cultivation must be limited to the "wine climate". Especially for shorter cultivation periods, the temperature requirements are met in large areas of Germany – only heavy, less air-permeable soils are ruled out. The water requirements and consumption of lucerne are high. However, it can cover its needs from deep layers (1.5 metres deep and more) if the soil is root-permeable after the sowing year. If the juvenile stage, which is sensitive to drought due to the still shallow root depth, is mastered, drought periods can then be withstood well. Frost hardiness is usually sufficient, apart from the juvenile stage. Late and variable frosts, combined with freezing of the soil, can cause damage. Poor vigour after winter is often the result of too much, especially too deep, cultivation in the previous year. Base-rich, loamy sands, sandy loams and well-aerated, loose soils that allow deep rooting are particularly suitable. Light sand is also tolerated without difficulty if clay-rich layers are present in the subsoil. In fissured limestone, which allows the roots to penetrate, even a shallow crumb depth is no obstacle to cultivation. Dense, cold, wet and acidic soils rule out lucerne cultivation. Stagnant moisture or impenetrable soil layers are particularly disadvantageous for lucerne. The soil must be at least as deep as possible. Despite the high lime requirement, the neutral range is sufficient for favourable growth. With its deep, intensive rooting, lucerne can obtain lime from a large volume of soil. Lucerne is not a pronounced lime plant. However, it responds positively to the warming, loosening and neutralising effect of lime.

3.1.2.2 Red clover (*Trifolium pratense* L.)

Although first discovered by Carl von Linné in meadows – which are rather extensive from today's perspective – this species is now used as "field red clover". Red clover forms a very deep, nutrient-storing taproot with numerous branching secondary roots in the year it is sown. In late autumn, the root head with the basal buds is pulled into the soil by shortening the roots, thus providing a certain degree of protection against the cold.

In spring, the shoots emerge directly at the soil surface, in contrast to lucerne, which grows its shoots higher up. Red clover is therefore less sensitive to grazing and deep cutting than lucerne, but is still much more sensitive than white clover.

The most important grouping of varieties is the number of *chromosome sets* (ploidy level) of the varieties. A distinction is made between *diploid* (= normal set of chromosomes) and *tetraploid* (= doubled set of chromosomes) varieties.

In southern Germany, the best diploid varieties achieve yields that correspond to the average of the tetraploid varieties in a multi-year, multi-site comparison. On average, the tetraploid varieties are more resistant to clover canker (*Sclerotinia trifoliorum* Erikss.) than the diploid varieties. The same applies to the southern stem borer (*Colletotrichum trifolii* Bain & Essary). However, individual diploid varieties have a comparable level of resilience. Tetraploid varieties have a 1.5% higher water content than diploid varieties at the same stage of development. This does not have a negative effect on green feeding. Slightly longer wilting times are required for preservation. The use of a conditioner can therefore make more sense here. The seed of tetraploid varieties is usually somewhat more expensive than that of diploid varieties. The reason for this is not the actual seed yield potential, but the higher risk of failure of tetraploid varieties during propagation. The mechanisms that lead to this are not yet fully understood.

Red clover is the most versatile and in Germany the most important clover species in field forage production. Red clover is most important in *perennial main crop cultivation* with a sowing year and a utilisation year. Two main crop years are possible depending on the persistence of the variety. Three productive main crop years are still rare. The proportion of pure red clover in total clover and clover-grass cultivation in Germany is less than 10%. Propagation cultivation differs from the usual consumer cultivation by significantly lower sowing rates (6–8 kg/ha compared to 16–20 kg/ha). In addition, red clover is used in small proportions as a mixture partner in meadow seeding.

It thrives best in a cool, temperate climate with medium humidity. A continental climate with long periods of drought in summer and long periods of heavy snow in winter makes it unsuitable for cultivation. However, red clover can survive short dry periods well due to its deep roots. Here it is often better than the grasses with which it is grown. The soil requirements are not strictly limited. From loamy sand to loam and clay, red clover tolerates all soils well. If the soil is light, the precipitation requirements are higher. Humus-poor, very light sands and acidic soils are not suitable for cultivation. Red clover copes much better than lucerne with subsoil deficiencies – high groundwater levels, soil compaction. Although red clover loves good soil contact, deep, thorough tillage does not make red clover cultivation unsafe.

Clover-grass mixtures can also be left standing for one year longer than the usual utilisation period of sowing year and utilisation year. On average, a drop in yield of less than 15% can be expected from the first to the second year. The more winter-hardy varieties currently available allow this longer-term utilisation. Experience has shown that this is only not advisable in locations with a high risk of red clover canker.

3.1.2.3 White clover (*Trifolium repens* L.)

The taproot of white clover is weaker than that of red clover and has more branches. The secondary roots grow well through the topsoil. The above-ground creeping shoots are characteristic of white clover. They are more or less strongly developed depending on the variety. These runners quickly close gaps and evenly shade the soil. The inflorescences are white, sometimes with a reddish tinge. The leaves are usually marked with a light-coloured triangle. White clover is used to a small to medium extent in perennial clover-grass cultivation. Here it is used less as a yield component and more to ensure the most comprehensive ground cover possible.

As a clover species that can be grazed and is grateful for grazing, the main use of white clover in permanent grassland is sowing in a mixture with grasses for pasture utilisation.

White clover is also used in meadow seeding with temporary grazing on heavy, moist soils. It is less important as an undersow in intercropping. The advantage here is that it remains well below the cover crop. Pure sowing only occurs in intercropping as green manure. White clover is characterised by its great adaptability. It thrives on all soils, with the exception of poor sandy soils and wet, cold clay soils. Its precipitation requirements are lower than those of red clover. However, moisture is necessary for good mass formation. Frost resistance is high. White clover is a pronounced light plant.

3.1.2.4 Berseem clover (*Trifolium alexandrinum* L.)

Berseem clover is not hardy, so it is only used as a main crop in summer cultivation. It is also used as a freezing catch crop. Seed production takes place in warmer EU countries such as Italy. There is no propagation in Germany. The thousand grain weight (TGW) is 2.7 to 3.2 g. A TWG of over 3.5 g is always a single-cut variety that is unsuitable for cultivation as a main crop. Single-cut varieties without regrowth are important for intercropping. Due to the rapidly increasing crude fibre content, the first cut should be taken at the latest at the beginning of flowering.

Berseem clover thrives best on light to medium-heavy, nutrient-rich soils in good condition. It is important that the soil is not acidic. It is grateful for regular, plentiful rainfall. In areas with over 800 mm annual precipitation, Berseem clover outperforms Persian clover in terms of yield. In dry years, both fail. They then often only produce 1 or 2 cuts. The heat requirements for juvenile development are relatively high.

3.1.2.5 Bastard clover or Alsike clover (*Trifolium hybridum* L.)

The rooting of Alsike clover is similar to red clover, but the taproot goes less deep and the lateral roots are shallower. In terms of flower colour, it occupies a middle position between red and white clover. It usually flowers in pink, with transitions to white and red. It is less productive than red clover due to the low level of cultivation in recent decades.

In red clover-unsafe, particularly moist locations, bastard clover is used as a mixing partner for clover grass and meadow seeding. Pure sowing hardly ever occurs, as pure bastard clover does not like to be eaten. It is very resistant to unfavourable soil and climatic conditions and tolerates moisture and cold better than red clover. Bastard clover thrives on all soils, even on wet and cold clay and peat soils. It also has low requirements in terms of lime content. Only dry sandy soil is not suitable for cultivation. Due to its low productivity and persistence, it is only of minor importance in years with a good seed supply of red clover.

3.1.2.6 Persian clover (*Trifolium resupinatum* L.)

Like Berseem clover, Persian clover is only suitable for summer annual cultivation. The propagation areas overlap. The TWG is between 1.2 and 1.4 g. The leaf shape is similar to clover, the stem is thick but hollow. Compared to Berseem clover, the stalk becomes woody less quickly. It can be utilised until full flowering without any loss of quality. It is mainly used in intercropping.

3.1.2.7 Bird's-foot trefoil (*Lotus corniculatus* L.)

Horned clover is also known as horned clover. It does not belong to the genus *Trifolium* and is therefore not related to white or red clover. Horned clover is a deep-rooting plant with a very strong tillering capacity. In the cultivated form it is perennial to perennial. Three to seven yellow to reddish-yellow butterfly flowers stand together in the form of a loose head. It takes its name from the horn-like curved pods. Bird's-foot trefoil has a certain importance for perennial clover-grass mixtures in dry locations, as well as in meadow seeding. Pure sowing is out of the question due to its slightly bitter flavour. Although it also thrives well in humid climates, it is primarily a plant of moderately dry areas. With a sufficient lime content, all soil types from heavy shell limestone to sandy and boggy soils are suitable. Other clovers are better on deep soils. Bird's-foot trefoil is superior on poor soils.

3.1.2.8 Crimson clover (*Trifolium incarnatum* L.)

The flowers in the elongated flower cluster are a deep red colour. The strongly branched taproot only penetrates shallowly into the soil. The dense hairiness of the leaves is remarkable. The light-coloured leaf markings of other clovers are missing. The value of crimson clover in winter intercropping is based on its rapid development in early spring. The first cut is 14 days before the red clover. Crimson clover is used in winter intercropping, e.g. with perennial ryegrass as a mixture partner. Due to the sparse regrowth, only one high-yielding cut can be expected. Its importance in practice is low today and lies, for example, in the fact that in Germany it is part of the so-called "Landsberg mixture". (21 kg/ha shaggy vetch [*Vicia villosa* L.], 9 kg/ha crimson clover (*Trifolium incarnatum* L.) and 30 kg/ha perennial ryegrass [*Lolium multiflorum* L.]). The typical weather conditions in this period are relevant for crimson clover as a winter catch crop. It does not survive hard, long, cold, snowy winters well. However, thinning does not have any serious consequences due to intercropping, as the partners fill in the gaps. It is grateful for well-drained, medium to light soil in good cultivation and lime condition.

3.1.2.9 Sainfoin (*Onobrychis viciifolia* Scop.)

It has a strong taproot that also penetrates stony ground. It has an elongated, racemose flower whose individual florets are light pink in colour. There are single- and multi-flowered forms. The winter hardiness of the multi-flowered form is lower than that of the single-flowered form. The sainfoin is only of minor importance as a mixture partner of perennial clover grass ("Jurassic clover grass") on shallow calcareous weathered soils. This species is currently attracting more attention in research, as its tannin content is of interest in forage-based rations in the field of animal nutrition.

Dry locations, due to the soil or rainfall, are a prerequisite for the cultivation of sainfoin. The drier the soil, the wetter the climate can be and vice versa. The heat requirements are lower than for lucerne.

References

Bayerischen Staatsministerium für Ernährung, Landwirtschaft und Forsten 2025 Bekanntmachung vom 13. Februar 2015 Az. L2-7364-1/6: Richtlinie zur Förderung von Umstrukturierungsmaßnahmen zur Verbesserung und Stärkung der Wirtschaftlichkeit von Grünfuttertrocknungsbetrieben (FuTrR).

Bellof G., Weindl P.A., Weindl P.N., 2021: Grünleguminosen als Eiweiß- und Raufuttermittel in der ökologischen Geflügel- und Schweinefütterung (GRÜNLEGUM). Schlussbericht BÖLN Projekt FKZ 2815OE039, 2815OE077, 2815OE079, 2815OE076, 2815OE073. https://orgprints.org/id/eprint/42739.

BLE 2024 https://www.bmel.de/DE/themen/landwirtschaft/pflanzenbau/ackerbau/eiweisspflanzenstrategie.html.

BMEL, 2012: Bundeslandwirtschaftsministerium fördert den Anbau von heimischen Eiweisspflanzen. agrar-PR (Download 14.8.24).

DeStatis, 2023: Wachstum und Ernte – Feldfrüchte – Fachserie 3 Reihe 3.2.1–16/2022 (Download 14.8.24).
Elgersma A., Søegaard K., 2016: Effects of species diversity on seasonal variation in herbage yield and nutritive value of seven binary grass-legume mixtures and pure grass under cutting, European journal of agronomy, Volume 78, Pages 73–83, ISSN 1161-0301, https://doi.org/10.1016/j.eja.2016.04.011.
Foster A., Vera C.L., Malhi S.S., Clarke F.R., 2014. Forage yield of simple and complex grass–legume mixtures under two management strategies. Canadian journal of plant science. 94(1): 41–50. https://doi.org/10.4141/cjps2013-095.
LegumHub, 2024: https://www.legumehub.eu (Download 14.8.24).
Lindermayer H., 2014: Wenn nicht Übersee-Soja, was dann? Zwischenbilanz und Ausblick der Eiweißstrategie, Tagungsband 13. BOKU-Symposium TIERERNÄHRUNG Wertvolle Pflanzenstoffe für die Tierernährung: Perspektiven und Entwicklungen 29. April 2014, 32–39, ISBN 978-3-900932-16-9.
Sleugh B., Moore K.J., George J.R., Brummer E.C., 2000: Binary legume–grass mixtures improve forage yield, quality, and seasonal distribution. Agron. J., 92: 24–29. https://doi.org/10.2134/agronj2000.92124x.
Suter M., Connolly J., Finn J.A., Loges R., Kirwan L., Sebastià M.-T., Lüscher A., 2015: Nitrogen yield advantage from grass–legume mixtures is robust over a wide range of legume proportions and environmental conditions. Glob change biol, 21: 2424–2438. https://doi.org/10.1111/gcb.12880.

3.2 Harvesting and preservation methods for obtaining protein-rich small-grain legumes products

Gerhard Bellof

3.2.1 *Introduction*

Reducing the use of feedstuffs that compete with humans for food is of central importance for greater sustainability in livestock farming (Eisler *et al.*, 2014). The protein of the soya extraction meal (48% CP) used in large proportions in pig and poultry feed, for example, has a human nutritional value of 90%. Compared to ruminants, which can digest very fibrous – and therefore inedible for humans – plant material thanks to their forestomach system, the digestive system of monogastric animals is similar to that of humans. In terms of nutrient composition, the requirements for pig and poultry feed are therefore comparable to those of humans. The proportion of protein in broiler rations that is potentially suitable for human consumption is 46% (Ertl *et al.*, 2016). Calculations of the food conversion efficiency for protein (human-nutritional output/human-nutritional input) show that this is only 0.52 for broiler chickens. The use of small-grain legumes as a source of protein and

TABLE 3.2.1-1 Ingredients (g/kg DM) in lucerne (*Medicago sativa L.*), lucerne silage and lucerne leaves

Feed stuff	CP	CF	Ash	Lysine	Methionine	Cysteine
Lucerne Whole plant, hot-air dried	243	299	106	13.1	3.7	2.5
Luzerne Whole plant, ensiled	226	296	110	13.2	3.4	1.1
Lucerne Leaves, hot-air dried	325	163	119	18.5	5.4	4.1

BELLOF (2017), UNPUBLISHED

amino acids for pigs and poultry could therefore reduce the proportion of feed that competes with human nutrition and thus improve the food conversion efficiency for pork and poultry meat.

Hoischen-Taubner and Sundrum (2016) tested exemplary selected varieties of the legume species lucerne, red clover, white clover, incarnate clover and Persian clover. In addition to lucerne, white clover in particular was characterised by a high crude protein content (g/kg DM) in the whole plant. White clover also had slightly higher methionine and lysine contents than red clover, but lucerne and red clover had the highest contents (g/kg DM) in the leaf mass. Both in the whole plant and in the leaf mass, white clover had higher in vitro precaecal digestibility values for methionine and lysine than lucerne and red clover. However, the latter achieved the highest crude protein yields (dt/ha) for whole plants and leaf mass (Hoischen-Taubner and Sundrum, 2016). Since white clover is also difficult to harvest due to its low growth height, lucerne and red clover appear to be more suitable for cutting young growth for silage preparation or for separation into leaves and stems. The crops listed in table 3.2.1-1 for various lucerne products illustrate these statements.

3.2.2 *Harvesting and preservation methods by comparison*

As early as 1980, a prototype was developed in the former German Democratic Republic that stripped and harvested lucerne leaves from the crop using rotating tines (Schmidt, 1997). The remaining stems were then cut off in a further step. A proportion of up to 15% stems in the harvested crop was tolerated, as these stems, which mostly originate from the upper parts of the plant, had a comparatively low crude fibre content (Schmidt, 1997). Further developments of a lucerne (alfalfa) leaf harvesting technique were advanced by Schinners *et al.* (2007) in the USA. This was a modified bean harvester with a scraper rotor. With this technique, up to 89.5% leaf content in the harvested crop was achieved. A 38.5% higher leaf yield was achieved compared to harvesting the lucerne as a whole plant. Lucerne leaves are higher in CP and lower in CF compared to the whole lucerne plant (table 3.2.2-1). The stems remaining in the field with residual leaves were chopped and harvested in a separate operation. In France, prototypes have recently been developed by the company Alf'ing-Trust'ing (44300 Nantes, France) for harvesting lucerne leaf mass from the standing crop by stripping it from the stalk using a finger roller.

Compared to leaf-stem separation directly in the field, since 2013 Qualitätstrocknung Nordbayern eG (DE-91710 Gunzenhausen) has had the option – after hot-air drying of the whole plants – of stationary extraction of the lucerne leaves using a sieving process at its site in Wechingen.

As part of the "Protein Regional – Lucerne-Clover" project, the leaves of small-grain legumes are harvested using a combine harvester converted at the TU Dresden. The leaves are separated from the stems during the threshing process of the wilted whole plants (lucerne at least 65% DM). This has the advantage that the harvested leaf material can already be stored. In order to lose as few leaves as possible during the relatively long wilting process in the field, it is recommended that the lucerne is placed in a wide swath during mowing and that the swath is gently turned once.

In addition to the technical separation of leaf mass and stems in small-grain legumes, Edwards *et al.* (1979) carried out the first trials to harvest the upper parts of the lucerne plant separately. A forage harvester with a 3.6 m wide header was used. This was raised during the first pass in order to harvest only the top 50% of the lucerne plant. The upper section of the plant, known as the lucerne tips, was then used as raw material for the production of a leaf protein concentrate and had a crude protein content that was 4.8% and 11.9% higher than the whole lucerne plants and the remaining lucerne stems respectively. In 2019, an innovative machine from Zürn Harvesting GmbH & Co. KG (DE-74214 Schöntal-Westernhausen) called "Top Cut Collect" was introduced. This was originally developed for mechanical weed control in the event of resistance problems or chemical-free arable farming strategies. The horizontal double-blade mower with reel, cross conveyor belt and a collection bunker (volume approx. 7 m^3) is very well suited for a "high cut", in which the plant tips and thus the plant parts with a higher protein content can be harvested separately, without separating the leaves and stems. The aim here is to achieve a cutting height of 40%. This means that if the lucerne plants grow to a height of 70 cm, the top 30 cm are harvested as lucerne tips.

Another technology for harvesting lucerne tips is the double-blade mower combination from BB-Umwelttechnik (DE-87672 Roßhaupten). This technology uses high skids to mow about 20% of the growth height (from the top) of the lucerne. The cut material is laid wide on the stubble, then raked and loaded with a loader wagon. This method offers advantages as it protects the plants and thus reduces harvest losses. The high skids prevent the lucerne plants from being driven over and pressed to the ground, resulting in a significantly better yield.

FIGURE 3.2.2-1 Harvesting lucerne plant tips by engine „Top cut collect“
PHOTO: © J. MAXA, LFL BAYERN

FIGURE 3.2.2-2 Harvesting lucerne plant tips by engine „BB-Umwelttechnik“
PHOTO: © J. MAXA, LFL BAYERN

As part of the interdisciplinary research project "GRUENLEGUM" (Bellof *et al.*, 2021), various of the described advertising methods were tested comparatively. The methods used to harvest lucerne or red clover can be outlined as follows:

> **Method 1 (M1):** Harvesting in the field, separation of leaf mass and stems by stripping the leaves in the standing crop and subsequent hot air drying.
> **Method 2 (M2):** Mowing and harvesting of the small-grain legumes as a whole plant with subsequent hot air drying, separation of leaf mass and stems by means of chopping and subsequent technical separation in a sorting plant.
> **Method 3 (M3):** Mowing and wilting (at least 60% DM) of the small-grain legumes in the field, drying in a hay ventilation system and subsequent separation of leaf mass and stems as in V2.
> **Method 4 (M4):** Mowing and wilting of lucerne (at least 65% DM) in the field with subsequent separation of leaf mass and stems using an adapted combine harvester.
> **Method 5 (M5):** Mowing the plant tips of lucerne with a high protein content as an alternative to leaf/stem separation, preservation of the lucerne tips by hot air drying.

Subsequent harvesting of lucerne residues (stems) and preservation by hot air drying.

The main results of the method comparison for methods 1 to 3 – separately for lucerne and red clover – are shown in tables 3.2.2-2 and 3.2.2-3.

TABLE 3.2.2-2 Total yield (DM), content of leaves and CP (per cut) for products of lucerne and red clover (free field) for methods 1–3 (means and standard deviation)

Method	Lucerne			Red clover		
	Yield [dt DM/ha]	Leave portion [%]	CP [% DM]	Yield [dt DM/ha]	Leave portion [%]	CP [% DM]
M1 leave	11.8 (1.1)	73 (3)	26.5 (1.6)	9.9 (1.0)	82 (2)	26.9 (1.3)
M1 stem	10.7 (1.2)	21 (4)	15.1 (2.1)	11.1 (1.9)	27 (3)	16.5 (1.8)
M1 total	22.5 (2.0)	48 (4)	–	21.0 (2.7)	55 (4)	–
M2	34.0 (2.9)	41 (2)	20.4 (1.4)	32.9 (3.6)	47 (3)	20.3 (1.1)
M3	29.8 (2.2)	32 (2)	18.3 (1.9)	30.8 (2.7)	38 (3)	18.9 (1.2)

MAXA *ET AL.* (2021), MODIFIED

With method 2, the highest yields (dt DM/ha) are achieved for both lucerne and red clover. However, the leaf/stem separation is less successful than for method 1. The resulting significantly higher leaf content for both legume species leads to a higher crude protein content in the "leaves" fraction. Method 3 performs worst in this comparison (table 3.2.2.2). Thus, the stationary leaf/stem separation using sieving seems less suitable, as the crude protein content is only slightly higher than the initial value of the whole plant.

The amino acid contents of lucerne and red clover products shown in table 3.2.2-3 reflect the different leaf proportions and the resulting different crude protein contents. The differences between the methods are greater than between the two legume species.

Important results of the comparison of methods carried out with lucerne only are shown for methods 4 and 5 in tables 3.2.2-4 and 3.2.2-5.

The annual comparison illustrates the large fluctuations in the dry matter yield of lucerne. Harvesting method V4 (combine harvester) shows advantages over the high-cut method (V5) in terms of total yield and crude protein yield (dt DM/ha). In contrast, the plant tips harvested with the V5 method have higher CP contents than the leaves (V4). The amino acid contents shown in table 3.2.2-5 reflect this fact exactly.

After analysing further data from small plot trials Maxa *et al.* (2021) draw the following conclusions for the 'harvesting of plant tips' method for the years 2019–2020:

The annual yield of lucerne tips at a cutting height of 40% is up to 57 dt of dry matter per ha, with an average crude protein content of 27% in the dry matter. This results in a crude protein yield of approx. 15 dt per ha. The crude protein yield is thus significantly higher than for domestic soya beans (assumed: 30 dt DM/ha; 38% CP in the DM).

TABLE 3.2.2-3 Average content of Lysine (Lys), Methionine (Met), Cysteine (Cys) and Threonine (Thr) for products of lucerne and red clover (free field) for methods 1–3 (g /kg DM)

Method	**Lucerne**				**Red clover**			
	Lys	**Met**	**Cys**	**Thr**	**Lys**	**Met**	**Cys**	**Thr**
M1 leave	16.3	4.5	3.4	12.4	16.1	4.3	2.7	13.0
M1 stems	8.3	2.2	1.9	6.4	8.9	2.3	1.6	7.1
M2	11.7	3.1	2.6	9.0	11.4	3.0	2.1	9.0
M3	10.2	2.7	2.3	8.1	10.2	2.7	1.9	8.2

MAXA *ET AL.* (2021), MODIFIED

TABLE 3.2.2-4 Average yield (DM), content of leaves and CP (per cut) for lucerne products (free field) for methods 4 and 5 (means and standard deviation)

Method	Harvest 2020 (1. cut)			Harvest 2020 (2. cut)		
	Yield [dt DM/ha]	Leave portion [%]	CP [% DM]	Yield [dt DM/ha]	Leave portion [%]	CP [% DM]
M4 Leaves	12.2 (3.1)	81 (3)	26.0	20.0 (2.1)	56 (4)	25.6
M4 Stems	7.4 (2.3)	0 (4)	15.2	15.2 (1.7)	0 (4)	19.3
M5 Plant tips	4.5 (1.6)	78 (9)	29.2	7.3 (2.8)	86 (5)	31.8
M5 Plant rest	15.6 (1.5)	30 (6)	18.4	25.1 (2.6)	13 (3)	19.3

MAXA *ET AL.* (2021), MODIFIED

TABLE 3.2.2-5 Average content of Lysine (Lys), Methionine (Met), Cysteine (Cys) and Threonine (Thr) for products of lucerne (free field) for methods 4 and 5 (g/kg DM)

Method	Harvest 2020 (1. cut)				Harvest 2021 (1. cut)			
	Lys	Met	Cys	Thr	Lys	Met	Cys	Thr
M4 leaves	13.2	3.8	3.0	11.1	12.9	3.4	2.9	11.0
M5 plant tips	15.7	4.3	3.3	12.6	16.5	4.6	3.6	13.3

MAXA *ET AL.* (2021), COMPLETED

The separately harvested plant residues had an average CP content of 18.0% (in the DM). The crude fibre content (CF) averaged 30.9% (in the DM). The stem-emphasised plant residues can therefore be considered as a possible structural feed for ruminants.

Damborg *et al.* (2020) describe a completely different technological process approach for obtaining high-quality, protein-rich products from small-grain legumes. A screw press is used to separate the plant sap (green juice) from the solid components of the plant (pulp). The simplified process diagram is shown in figure 3.2.2-3. According to studies by Damborg *et al.* (2020), the pressed juice obtained from lucerne contains 323 g CP per kg DM, while the pulp contains 184 g CP per kg

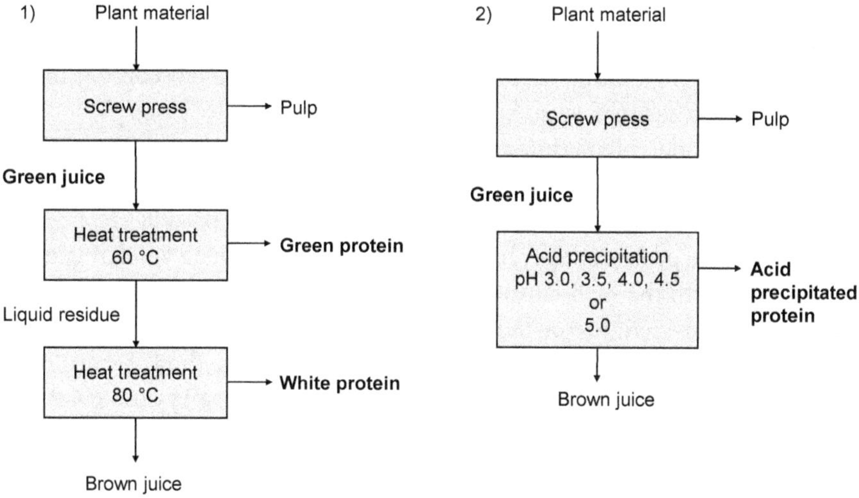

FIGURE 3.2.2-3 Green protein extraction – process overview. The method of green juice extraction is identical, but the precipitation of protein from the juice is performed by either a two-step heat precipitation (1) or acidic precipitation (2). DAMBORG *ET AL.* (2020), MODIFIED

TABLE 3.2.2-6 Distribution of DM and CP in the plant and throughout the juice extraction

Item	Legume	Plant, whole	Juice	Pulp	Waste
		g/kg or g/kg DM	g/kg DM	g/kg DM	g/kg DM
DM	Lucerne	196	55.9	123.5	16.9
	Red clover	166	49.5	91.5	25.1
CP	Lucerne	217	92.4	116.3	8.2
	Red clover	205	71.8	108.2	24.8

DAMBORG *ET AL.* (2020), MODIFIED

DM (table 3.2.2-6). The pressed juice can be further upgraded by means of heat or acid treatment and thus achieve high protein contents (lucerne: 405 g CP/kg DM).

Further investigations by Renaudeau *et al.* (2022) with heat-treated lucerne press juice resulted in an even higher crude protein content (537 g CP/kg DM) with correspondingly high levels of essential amino acids (e.g. 35.7 g lysine/kg DM), which reach the level of soya extraction meal (48% CP). The resulting product is therefore suitable for feeding monogastric animals. The fibre-rich pulp, on the other hand, should be fed to ruminants.

However, as shown in table 3.2.2-6, the dry matter content of the pressed juice is only 29% (lucerne) or 30% (red clover) of the total plant. Since the pulp – as already shown – still has a high crude protein content, the crude protein content of the pressed juice is only 43% (lucerne) and 35% (red clover).

The following model calculation should further clarify the situation. The pressed juice of lucerne contains 92.4 g CP per kg dry matter (table 3.2.2-6). Assuming an annual yield of 100 dt dry matter lucerne per ha, the crude protein yield of the pressed juice produced is 9.2 dt per ha. In contrast, a crude protein yield of 15 dt/ha can be achieved with the high-cutting method described above for the extraction of lucerne tips. Consideration of the possible yields of amino acids from pressed juice extraction improves the result slightly in favour of this method. According to studies by Stødkildet *et al.* (2019), lucerne press juice has 6.62 g Lys/100 g CP and 1.94 g Met/100 g CP. Using the above data, this results in the following yields: 61 kg lysine and 18 kg methionine per ha. The comparative calculation for the lucerne tips results in 90 kg lysine and 25 kg methionine/ha (assuming amino acid contents from table 3.2.2.5).

Brooke *et al.* (2023) compared the two methods of leaf/stem separation and pressed juice extraction. The results are shown using the example of the red clover plant (location Sweden). The annual yield from three cuts for the entire plant was 16.8 dt dry matter per ha with an average of 25% CP in the DM. This yield was distributed in a ratio of 43% to 57% between the leaves and stems respectively. For pressed juice extraction, the ratio was 35% juice to 65% pulp. The leaves had a higher CP content than the pressed juice (32% CP in the DM versus 23% CP in the DM). Maxa *et al.* (2021) determined significantly higher dry matter yields for a location in southern Germany (table 3.2.2.2). There were also considerable differences in the leaf to stem ratio (82% to 27%). In contrast, the average CP content in the red clover leaves was 26% (in the DM), which was lower than for the Swedish leaf material.

References

Bellof G., Weindl P.A., Weindl P.N., 2021: Grünleguminosen als Eiweiß- und Raufuttermittel in der ökologischen Geflügel- und Schweinefütterung (GRÜNLEGUM). Schlussbericht BÖLN-Projekt. https://orgprints.org/id/eprint/42739.

Brooke M., Steffen A., Forkman J., Parsons D., 2023: Production and nutrient composition of forage legume fractions produced by juicing and leaf stripping. Acta agriculturare Scandinavia, section B – soil & plant science, 73, 199–212. doi: 10.1080/09064710.2023.2265147.

Damborg V.K., Jensen S.K., Weisbjerg M.R., Adamsen A.P., Stødkilde L., 2020: Screw-pressed fractions from green forages as animal feed: chemical composition and mass balances. Anim feed sci technol., 261, 114401. doi:10.1016/j.anifeedsci.2020.114401.

Edwards R.H., Knuckles B.E., Miller R.E., Currence D.H., De Fremery D., Kohler G.O., 1979: Use of leaf enriching harvesting methods to increase the yield of leaf protein concentrate from lucerne. Journal of the science of food and agriculture, 30, 558–565.

Eisler M.C., Lee M.R.F., Tarlton J.F., Martin G.B., Beddington J., Dungait J.A.J., Greathead H., Liu J., Mathew S., Miller H., Misselbrook T., Murray P., Vinod V.K., van Saun R., Winter M., 2014: Agriculture: steps to sustainable livestock. Nature, 507, 32–34. doi:10.1038/507032a.

Ertl P., Steinwidder A., Schönauer M., Krimberger K., Knaus W., Zollitsch W., 2016: Net food production of different livestock: a national analysis for Austria including relative occupation of different land categories. Die Bodenkultur: journal of land management, food and environment, 67, 91–103. doi: 10.1515/boku-2016-0009.

Hoischen-Taubner S., Sundrum A., 2016: Ermittlung des Futterwertes und der Verdaulichkeiten der Blattmassen von Luzerne und Perserklee. Retrieved from http://www.orgprints.org/30426.

Maxa J., Liebhardt S., Thurner S., 2021: Untersuchungen zur Blatt-Stängeltrennung und Konservierung von Grünleguminosen (Arbeitspaket Verfahrenstechnik). Editors: Bellof G., Weindl P.A., Weindl P.N., 2021: Grünleguminosen als Eiweiß- und Raufuttermittel in der ökologischen Geflügel- und Schweinefütterung (GRÜNLEGUM). Schlussbericht BÖLN Projekt FKZ 2815OE039, 2815OE077, 2815OE079, 2815OE076, 2815OE073. https://orgprints.org/id/eprint/42739.

Renaudeau D., Jensen S.K., Ambye-Jensen M., Adler S., Bani P., Juncker E., Stødkilde L., 2022: Nutritional values of forage-legume-based silages and protein concentrates for growing pigs. Animal, 16, 100572. doi: 10.1016/j.animal.2022.100572.

Schinners K.J., Herzmann M.E, Binversie B.N., Digman M.F., 2007: Harvest fractionation of alfalfa. Transactions of the ASABE 50, 713–718. doi: 10.13031/2013.23125.

Schmidt L., 1997: Separate harvest of leaves from the stems of lucerne stands. Forage and grassland management, Session 22.

Stødkildet L., Damborg V.K., Jørgensen H., Lærke H.N., Jensen S.K., 2019: Digestibility of fractionated green biomass as protein source for monogastric animals Animal, 13, 1817–1825. doi: 10.1017/s1751731119000156.

3.3 Antinutritive factors in small-grain legumes

Petra Weindl

Besides their very promising nutritive properties according protein and amino acid content (section 3.4), many small-grain legumes simultaneously contain very efficient defense compounds against herbivores or pathogens. These antinutritive factors can lead to detrimental effects regarding growth and performance, when incorporated to animal diets.

The most frequently mentioned phytochemicals in small-grain legumes are phytoestrogens, saponins and cyanogenic glucosides (Vetter, 2000; Carlsen a. Fomsgaard, 2008; Gierus *et al.*, 2012).

However, the probably most important antinutritive factors in small-grain legumes and especially in lucerne are the so called saponins (Sen *et al.*, 1998). Saponins are secondary metabolites, which can be found in plants and certain marine organisms, consisting of an isoprenoid core covalently linked to one or more sugar residues (Augustin *et al.*, 2011). In plants they serve as a natural defense system against for example herbivores, microbes or funghi.

Forage legumes are well known to contain saponins (Cheeke, 1983a). In particular those within the genus of snail clover (*Medicago* L.), exhibit a variety of bioactive and structurally diverse saponins (Oleszek and Jurzysta, 1986a; Oleszek *et al.*, 1990; Bialy *et al.*, 1999). But, also red and white clover contain saponin compounds (Oleszek and Jurzysta, 1986b; Sakamoto *et al.*, 1992; Carlsen and Formsgaard, 2008). A lot of research has already been carried out on lucerne saponins in the past.

3.3.1 *Saponins*

3.3.1.1 Saponins of the lucerne plant

Generally *Medicago sativa* L. saponins can be described as a complex mixture of mono- and bidesmosidic saponins with short- and long-chain saccharide residues (Shany *et al.*, 1970; Levy *et al.*, 1986; Oleszek and Jurzysta, 1986a; Massiot *et al.*, 1988; Oleszek *et al.*, 1990; Massiot *et al.*, 1991; Massiot *et al.*, 1992; Oleszek *et al.*, 1992; Tava *et al.*, 1993; Bialy *et al.*, 1999; Huhman and Sumner, 2002; Mazahery-Laghab *et al.*, 2010). The following six aglycones are predominant in *Medicago* sativa L.: Hederagenin, Bayogenin, Sapogenols E, Oleanolic acid, Medicagenic Acid and Zanhic Acid (Huhman and Sumner, 2002; Tava and Avato, 2006). Both leaves and roots of *Medicago sativa* L. contain monodesmosidic saponins (predominantly substituted at position 3), as well as bidesmosidic saponins (substitued at positions 3 and 28) with various monosaccharide units such as D-Glucose, L-Arabinose, D-Xylose, D-Galactose, D-Glucoronic Acid and L-Rhamnose (Sen *et al.*, 1998). Although saponins are present in leaves, roots, flowers and seeds of the lucerne plant, the concentration of saponins in aerial parts is higher for leaves compared to stems (Pedersen *et al.*, 1967).

3.3.1.2 Significance of saponins for monogastric nutrition

Due to their structure and amphiphilic nature, saponins often exhibit biological activity. The biological activity is determined on the one hand by the non-polar, cholesterol-like aglycone and on the other hand by the number and composition of sugar residues. In addition, the spatial structure does play a crucial role in biological activity of lucerne saponins (Cheeke, 1971; Sen *et al.*, 1998; Francis *et al.*, 2007). The primary antinutritive effects of leguminous saponins on monogastric animals, include reduced feed intake due to their characteristic bitter taste, growth depression, hemolysis, reduced nutrient absorption and increased cholesterol excretion Cheeke (1983a).

Both reduced feed intake and impaired nutrient absorption are primarily responsible for the documented negative effects on growth intensity in literature. Leamaster and Cheeke (1979) reported feed refusal in fattening pigs with a content of already 1% lucerne meal in the diet. Geese, turkeys, quails and chickens showed lower sensitivity to lucerne meal included in their diet (Cheeke *et al.*, 1983b). Geese preferred lucerne-free diets from a content of 2.5% lucerne meal, while quails and turkeys favored diets from 5% and chicken from 10%. It is assumed that the bitter taste of saponins leads to reduced feed intake (Leamaster and Cheeke, 1979; Cheeke *et al.*, 1983b).

Another antinutritive effect of saponins is their action within the gastrointestinal tract. Due to their chemical structure, saponins have detergent-like characteristics and can form complexes, especially with substances rich in lipids. The cholesterol-binding effect of saponins has been described in literature (Peterson, 1950; Gee and Johnson, 1988) and is often considered as a positive effect in human nutrition. Feeding saponin-containing feed like lucerne to broiler chicken could promote the intake of low-cholesterol animal products in human diets (Ponte *et al.*, 2004; Weltin *et al.*, 2014). However, saponins also interact with lipid-rich cell surface structures in both animal and human organisms. A reduction of cell surface tension leads to cell membrane damage, increased permeability and ultimately cell death. Through this mechanism, saponins are enabled to damage erythrocyte membranes, as observed in the hemolysis assay for detecting hemolytically active saponin compounds (Jurysta, 1979). In case of cell destruction in enterocytes, the damage of intestinal cell integrity allows the passage of large molecules through the intestinal barrier. For human nutrition this presents a risk of potential allergen transfer (Gee *et al.*, 1996; Gee *et al.*, 1997). For animal nutrition, however, it superficially impairs the absorption of essential macro and micronutrients and therefore again decreases performance (Southan *et al.*, 1988; Gee *et al.*, 1993; Gee *et al.*, 1996; Gee *et al.*, 1997). Pleger *et al.* (2021) for example reports a lower precaecal digestibility for amino acids when feeding lucerne and red clover – leaves to broilers. The formation of complexes between bile acids and saponins is also a very effective protective mechanism to prevent the entry of antinutritive substances into the bloodstream up to certain concentrations (Gee a. Johnson, 1988). In vitro studies on prepared rat small intestines showed that saponins affect the transmural potential difference of mucosal cells, preventing the absorption of certain nutrients from the diet (Johnson *et al.*, 1986; Gee and Johnson, 1988; Gee *et al.*, 1989; Oleszek *et al.*, 1994). The most relevant structure-dependent biological activities from the perspective of animal nutrition like bitter taste, reduction of the transmural potential difference and antimicrobial properties are summarized in table 3.3.1-1. Hemolytic activity is provided additionally.

The abovementioned results suggest that not only knowledge of the total content of saponins but also knowledge about the structure and biological activity of individual saponins in lucerne products is essential for the assessment and use of these feedstuffs.

TABLE. 3.3.1-1 Structure-dependent differences in the biological activity (sites of action in the animal organism) of selected saponins according Weindl *et al.* (2021)

Saponin	**Feed intake**	**Gastrointestinal tract**	**Gastrointestinal tract**	**Metabolism**
	Bitter taste, mucosal irritation	Reduction of the transmural potential difference (electrochem. gradient) of small intestine	Antimicrobial	Hemolytic
Saponines of the aglycone of medicagenic acid	yellow	yellow	red	red
Saponines of the aglycone of Zanhic acid	red	red	yellow	yellow
Saponines of the aglycone of Soyasapogenols	green	green	green	green

Legend: green – no impact; yellow – low impact; red – high impact.

3.3.1.3 Significance of saponins for ruminants

While in the past saponins have long been suspected of being responsible for bloat in ruminants (Cheeke, 1996) recent studies are more interested in the impact on the ruminal microbiota concerning the reduction of climate relevant emissions.

According to recent research, saponins are capable to effectively reduce methane emissions by especially affecting microorganisms involved in methane production, mostly protozoa. Kozlowksa *et al.* (2020) proved that the lucerne varieties "Kometa" and "Verko" effectively mitigate methane production in vitro, without affecting the basic fermentation parameters. Wina *et al.* (2005), however report of only a transient antiprotozoal effect, when tested in in vivo experiments. There seems to be an adaptation of the ruminal microflora towards saponins (Wina *et al.*, 2005).

3.3.1.4 Impacts on the saponin content of lucerne as a forage plant

As the saponin content of lucerne is determined by several impacts such as variety, cultivation period and climate factors (Tava *et al.*, 1999; Pecetti *et al.*, 2006), the choice of variety and harvesting time can be decisive for the (anti)/-nutritive properties of small-grain legumes. In a large-scale research project called "GRÜNLEGUM" Weindl *et al.* (2021) investigated, the influence of variety, harvesting date and processing on the saponin content of lucerne. The significance of these factors for selected saponins is examined in more detail below. Table 3.3.1-2 illustrates the impact of variety on saponin composition in dried lucerne leaves. Differences between the varieties are in particular relevant for medicagenic and zanhic acid saponins, which are suspected to be the major interfering factors for animal nutrition (table 3.3.1-1). The variety "Sanditi" shows about two times higher medicagenic and approximately a half less zanhic acid compared with "Fleetwood". "Plato" lies between the aforementioned varieties (table 3.3.7-3).

The relevance of harvest time as well as the distribution of certain saponins within the lucerne plant is indicated in table 3.3.1-3. Leaves have consistently higher medicagenic and zanhic acid contents than stems. Additionally, there are clear differences in medicagenic and zanhic acid contents between harvesting dates. Zanhic acid contents are highest in June and July.

Another influencing factor is the processing method, which seems to affect the structure and quantity of saponins and therefore their biological activity. Cheeke *et al.* (1977) investigated lucerne plant material with high and low saponin contents under varying processing methods (freezing, freeze drying, air drying at 57 °C). The saponin effect was not related to temperature, when feeding those diets to rats. In contrast ensiling seems to be effective in restructuring and reducing saponins. Ensiling resulted in structural and quantitative changes of saponins of in total ten lucerne varieties Szumacher-Strabel *et al.* (2019). Kalac *et al.* (1996) compared saponin concentrations of leaves from two lucerne varieties ("Hodoninka", "Palava") in fresh material and under silage conditions. Besides significant differences of total saponin content between varieties as well as individual saponins, clear effects were observed due to the ensiling process. Through this process, certain

TABLE 3.3.1-2 Composition of selected saponins in dried lucerne leaves of different varieties (data in µg/g eq umbelliferone (LS means)) according Weindl *et al.* (2021)

Saponin	Sanditi	Plato	Fleetwood	*p*
Medicoside J	9.2^{a}	6.3^{b}	4.3^{b}	0.0371
Medicoside H	6.2^{a}	$5.2^{a,b}$	4.0^{b}	0.0013
Medicagenic acid	57.7^{a}	35.6^{b}	28.0^{b}	0.0096
Zanhic acid	5.4^{a}	6.9^{a}	10.8^{b}	0.0016

1 *p*: probability of error.
a–c Different superscripts indicate significant differences between the subclasses ($p \leq 0.05$)

TABLE 3.3.1-3 Influence of development stage on the contents of selected saponins in dried lucerne leaves and stems of the variety "Plato" (in µg/g eq umbelliferone) according Weindl *et al.* (2021)

Sampling date	Cut Nr.	Leaves		Stems	
		Zanhic-acids	Medicagenic acids	Zanhic acids	Medicagenic acids
19.04.18	1	1.7	141.9	0.8	83.7
26.04.18	1	4.5	166.1		
10.05.18	1	4.9	157.9		
24.05.18	2	4.9	167.1	0.9	63.5
31.05.18	2	15.9	299.8	5.5	156.7
07.06.18	2	12.3	221.3		
21.06.18	2	8.5	166.5		
05.07.18	3	9.2	154.8	1.7	68.8
25.07.18	3	10.9	191.1	4.2	123.3
16.08.18	3	3.2	72.7	1.8	60.3

saponins (including medicagenic acid) were degraded by up to 100%. However, Szumacher-Strabel *et al.* (2019) believe that the majority of these glycosides are chemically restructured during ensiling and therefore it is difficult to characterize silage quality only by determining the saponin content. This statement is confirmed by the findings reported by Weindl *et al.* (2021) using a semi-quantitative method for analysing saponins in four lucerne products. Known saponins from literature were identified via Time of Flight Liquid Chromatography (LC/TOF) and set in relation to the laboratory standard umbelliferone with a known concentration. The products analysed were: lucerne leaves, dried at high and low temperature, lucerne silage (whole plant (WP), harvested in 2017); and lucerne meal (WP) harvested in 2016. Table 3.3.1-3 lists the most dominant saponins in terms of concentration, found in the samples (> 50mg/kg eq umbelliferone). It has to be mentioned that the four lucerne products derived from two different varieties and different harvests (2016 and 2017). Thus, the results do allow certain statements about the influence of the saponin content (leaf/whole plant, dried/ensiled), but they do not represent a strategic investigation of the various influencing factors (Weindl *et al.*, 2021).

The contents of individual saponins vary strongly (table 3.3.1-4) between the different products. The analysed lucerne meal contains consistently low levels of saponins. This could be proven in an in vivo experiment by Pleger *et al.* (2018), feeding the here analysed lucerne meal to broilers. Body weight and digestibility were

TABLE 3.3.1-4 Results of saponin analysis of four lucerne products (> 50mg/kg eg umbelliferon) according Weindl *et al.* (2021)

Saponin	Aglycon	Lucerne leaves High temp.	Lucerne leaves Low temp.	Lucerne silage WP* 2017	Lucerne meal WP* 2016
		(µg/g eq umbelliferon)			
Dehydrosoyasaponin I	Soyasaponin	2.7	2.9	107.1	1.8
Hex-Hederagenin	Hederagenin	11.3	17.8	65.6	4.2
Dehydrosoyasaponin I	Soyasaponin	38.2	39.3	56.2	31.4
Medicagenic acid derived	Medicagenic acid	51.4	50.3	62.6	21.3
Hex-HexA-Aglycone A	Unknown	61.4	70.5	23.4	28.0
Azukisaponin II	Azukisaponin	62.7	58.6	111.4	34.2
Medicoside H	Medicagenic acid	84.2	58.5	208.9	27.9
Medicagenic acid 3-O-b-D-glucuronide	Medicagenic acid	89.3	87.9	242.1	13.0
3-glc-glc-28-ara-rha-medicagenic acid	Medicagenic acid	100.7	94.4	138.0	40.0
3-glcA-28-ara-rha-xyl-zanhic acid	Zanhic acid	155.4	117.9	17.8	12.8
HexA-dhex-pen-pen-pen-zanhic acid	Zanhic acid	199.8	175.4	12.4	47.9
Total saponin content (Of all 41 detected saponins)		1,160	1,022	1,318	368

* Whole plant.

generally on a good level and not impaired when feeding lucerne meal WP 2016. The feeding of lucerne silage WP 2017 did not show any detrimental effects during another in vivo experiment with broilers neither (Pleger *et al.*, 2021). However, feeding lucerne leaves low temp. led to severe performance depression in a third broiler trial by Pleger *et al.* (2020). Even though lucerne silage WP 2017 presented higher values for total saponin content and for some individual saponines compared to the lucerne leaves (table 3.3.1-4). The comparatively high contents of the aglycone zanhic acid in lucerne leaves (table 3.3.1-4) seem to play a crucial role for assessing the feed value of lucerne products. These results demonstrate, that the total saponin content does not allow a sufficient assessment for the use of lucerne products in feed. Individual saponins are crucial for the acceptance of lucerne products, which requires individual saponin identification and quantification.

In order to fully exploit the potential of small-grain legumes especially as protein alternatives it would be necessary to elucidate the structure and content of antinutritive saponins more profound. Especially restructuring processes in the course of processing (ensiling, drying) in combination with their biological activity would be of great interest here. Due to their diversity and various distributions in the plant, there are currently no routine analysis methods for detecting and quantifying lucerne or legume saponins. However, knowledge about the antinutritive value and the establishment of specific thresholds for various animal species would be essential for incorporating lucerne products in higher amounts to animal diets.

3.3.2 *Further antinutritive factors in small-grain legumes*

3.3.2.1 Phytoestrogens

Gierus *et al.* (2012) states that legumes in general are the primary source of phytoestrogens in animal nutrition. Phytoestrogens like isoflavones are predominant in red clover, whereas coumestans and lignans are mainly contained in white clover and lucerne (Gierus *et al.*, 2012). Their ability to be transferred into milk from a dairy cow diet gives these substances a special significance for human nutrition (Gierus *et al.*, 2012). Renaudeau *et al.* (2022) compared lucerne and red clover fresh products, silages and protein pastes in terms of known secondary compounds. The isoflavonoid contents detected in red clover showed higher concentrations of biochanine A, formonentin and genistein compared to lucerne. The highest levels of these isoflavonoids were detected in red clover fresh leave and protein paste. In contrast lucerne products (whole plant, leaves and silages) were higher in luteolin, tricin and apigenin (Renaudeau *et al.*, 2022).

3.3.2.2 Trypsin inhibitors (TI)

Trypsin inhibitors interact with the protein cleaving enzyme trypsin. Leading to an inhibition of the subsequent proteolytical activity cascade of further enzymes involved in protein degradation, initiated by trypsin (Stangl, 2011). TI were isolated from the lucerne plant (Brown and Ryan, 1984; Chien and Mitchell, 1970), showing the highest concentrations in leaves and increasing contents with maturation of the plant (Chang *et al.*, 1978). However, their nutritional effect on animals is unknown (Chang *et al.*, 1978). A low TI activity was also found in red clover seeds (Norioka *et al.*, 1988). In studies by Maliar *et al.* (2011), various forage plants, including Medicago sp. and Trifolium sp. were tested for proteinase inhibition. While the subgroups of Medicago representatives showed the lowest average TI activity, a subgroup of Trifolium representatives showed the highest average TI activity compared to black tea extract.

3.3.2.3 Tannins

Tannins belong to the group of polyphenols. Besides their positive effect of protecting proteins from degradation during the ensiling process (Albrecht and Muck, 1991), they reduce crude protein digestibility in monogastric animals (Jansman *et al.*, 1993; Ortiz *et al.*, 1993). This is due to their binding capacity to feed protein, endogenous protein and directly to the enzymes of the digestive tract (Van Leeumen *et al.*, 1995). In addition, studies have shown that tannins have toxic and growth-inhibiting effects in various animal species (Dollahite *et al.*, 1962; Glick a. Joslyn, 1970; Rayudu *et al.*, 1970). However, tannins are partially degraded during ensiling (Gefrom *et al.*, 2013; Lorenz *et al.*, 2010). Silages can therefore improve the amino acid supply for non-ruminants without presenting antinutritive properties (Kofahl, 2009). Tannins are also influenced by thermal processes (Serrano *et al.*, 2009). Red clover whole plant and leaves contain more condensed tannins compared to lucerne (Renaudeau *et al.*, 2022). However, as lucerne and red clover contain only low levels of condensed tannins (red clover: 0.27–1.7% DM, lucerne: 0.07–0.5% DM) compared to other small-grain legumes, (Barry and McNabb, 1999; Hamacher *et al.*, 2015) their significance as antinutritive factors can be probably considered as low.

3.3.2.4 Polyphenol oxidase (PPO)

PPO or tyrosinase is a copper-containing enzyme present in the thylakoids of chloroplasts. PPO catalyzes the oxidation of phenols to o-quinones using molecular oxygen (Gierus, 2009). However, these highly reactive molecules either polymerize to form melanins or form complexes with other nucleophilic molecules (e.g. proteins) (Gierus, 2009; Lee *et al.*, 2007; Lee *et al.*, 2004). This browning reaction is also observed in the harvest of red clover leaves, which is why Jones *et al.* (1995) concluded that PPO is present in red clover. Renaudeau *et al.* (2022) found that PPO activity was present in red clover and lucerne, with a higher activity for red clover. Protein degradation in red clover silage was 40% lower than in lucerne silage, despite the same dry matter content, pH value progression during ensiling and final pH value in the silage (Jones *et al.*, 1995). Consequently, PPO leads to lower proteolysis during ensiling in red clover compared to lucerne (Hatfield and Muck, 1999; Jones *et al.*, 1995; Lee *et al.*, 2004; Sullivan and Hatfield, 2006).

3.3.2.5 Salicylic acid

In addition, lucerne is known to contain a substance with pharmacological effects. Salicylic acid has blood thinning properties, increases toxin tolerance and cell wall stabilization (Pütz, 2014). Preissinger *et al.* (2016) conducted a feeding trial with weaned and non-tail-docked piglets offering lucerne hay, grass cobs and maize silage ad libitum. Significantly higher daily weight gains and a significant reduction of tail biting was achieved by piglets receiving lucerne compared to the the piglets of other groups.

References

Albrecht K.A., Muck R.E., 1991: Proteolysis in ensiled forage legumes that vary in tannin concentration. Crop sci., 31, 464–469.

Augustin J.M., Kuzina V., Andersen S.B., Bak S., 2011: Molecular activities, biosynthesis and evolution of triterpenoid saponins. Phytochem., 72, 435–457.

Barry T.N., McNabb W.C., 1999: The implications of condensed tannins on the nutritive value of temperate forages fed to ruminants. Br. J. Nutr., 81, 263–272.

Bialy Z., Jurzysta M., Oleszek W., Piacente S., Pizza C., 1999: Saponins in alfalfa (Medicago sativa L.) root and their structural elucidation. J. Agr. Food chem., 47, 3185–3192.

Brown W.E., Ryan C.A., 1984: Isolation and characterization of a wound-induced trypsininhibitor from alfalfa leaves. Biochem., 23, 3418–3422.

Carlsen S.C., Fomsgaard I.S., 2008: Biologically active secondary metabolites in white clover (Trifolium repens L.) – a review focusing on contents in the plant, plant–pest interactions and transformation. Chemoecology, 18, 129–170.

Chang, H.-Y., Reeck G.R., Mitchell H.L., (1978): Alfalfa trypsin inhibitor. J. Agr. Food chem., 26, 1463–1464.

Cheeke, P.R., 1971. Nutritional and physiological implications of saponins: a review. Can. J. Anim. Sci., 51, 621–632.

Cheeke P.R., Kinzell J.H., Pedersen M.W., 1977: Influence of saponins on alfalfa utilization by rats, rabbits and swine. J. Anim. Sci., 46, 476–481.

Cheeke P.R., 1983a: Biological properties and nutritional significance of legume saponins. Editors: Telek, L., Graham H.D. In: Leaf protein concentrates. The Avi Publishing Company, Westport, Connecticut.

Cheeke P.R., Powley J.S., Nakaue H.S., Arscott G.H., 1983b: Feed preference responses of several avian species fed alfalfa meal, high- and low-saponin alfalfa, and quinine sulfate. Can. J. Anim. Sci., 63, 707–710.

Cheeke P.R.,1996: Biological effects of feed and forage saponins and their impacts on animal production. Editors: Waller G.R., Yamasaki K. In: Saponins used in food and agriculture. 405, New York.

Chien T.F., Mitchell H.L., 1970: Purification of a trypsin inhibitor of alfalfa. Phytochemistry, 9, 717–720.

Dollahite J.W., Pigeon R.F., Camp B.J., 1962: The toxicity of gallic acid, pyrogallol, tannic acid, and Quercus havardi in the rabbit. Am. J. Vet. Res., 23, 1264–1267.

Francis G., Kerem Z., Makkar H.P.S., Becker K., 2007: The biological action of saponins in animal systems: a review. Brit. J. Nutr., 88, 587–605.

Gee J.M., Johnson I.T., 1988: Interactions between hemolytic saponins, bile salts and small intestinal mucosa in the rat. J. Nutr., 118, 1391–1397.

Gee J.M., Price K.R., Ridout C.L., Johnson I.T., Fenwick G.R., 1989: Effects of some purified saponins on transmural potential difference in mammalian small intestine. In Vitro toxicol., 3, 85–90.

Gee J.M., Price K.R., Ridout C.L., Wortley G.M., Hurrell R.F., Johnson I.T., 1993: Saponins of quinoa (Chenopodium quinoa): effects of processing on their abundance in quinoa products and their biological effects on intestinal mucosal tissue. J. Sci. Food agric., 63, 201–209.

Gee J.M., Wortley G.M., Johnson I.T., Price K.R., Rutten A.A.J.J.L., Houben G.F., Penninks A.H., 1996: Effects of saponins and glycoalkaloids on the permeability and viability of mammalian intestinal cells and on the integrity of tissue preparations In Vitro. In Vitro toxicol., 10, 117–128.

Gee J.M., Wal J.M., Miller K., Atkinson H., Grigoriadou F., Wijnands M.V.W., Penninks A.H., Wortley G., Johnson I.T., 1997: Effect of saponin on the transmucosal passage of ß-lactoglobulin across the proximal small intestine of normal and ß-lactoglobulin sensitised rats. Toxicol., 117, 219–228.

Gefrom A., Ott E.M., Hoedtke S., Zeyner A., 2013: Effect of ensiling moist field bean (Vicia faba), pea (Pisum sativum) and lupine (Lupinus spp.) grains on the contents of alkaloids, oligosaccharides and tannins. J. Anim. Physiol. An. N., 97, 1152–1160.

Gierus P.M., 2009: Bedeutung von sekundären Pflanzeninhaltsstoffen für die Proteinqualität von Futterpflanzen. 15. Alpenländisches Expertenforum – Grundfutterqualität: aktuelle Ergebnisse und zukünftige Entwicklungen, Download: 18. Juni 2020 URL: https://www.researchgate.net/profile/Martin-Gierus/publication/252931029_Bedeutung_sekundarer_Pflanzeninhaltsstoffe_fur_die_Bestimmung_der_Proteinqualitat/links/0deec51f53c8b53346000000/Bedeutung-sekundaerer-Pflanzeninhaltsstoffefuer-die-Bestimmung-der-Proteinqualitaet.pdf, 2009.

Gierus M., Koch M., Scholz H., 2012: Übertragung von Phytoöstrogenen aus Leguminosen in die Milch – Eine Betrachtung entlag der Wertschöpfungskette. In: Berichte über Landwirtschaft, Hrsg.: Bundesministerium für Ernährung, Landwirtschaft und Verbraucherschutz 90, 354–379. ISSN 0005-9080.

Glick Z., Joslyn M.A., 1970: Food intake depression and other metabolic effects of tannic acid in the rat. J. nutr., 100, 509–515.

Hamacher M., Loges R., Taube F., 2015: Zum Nutzungspotential alternativer Leguminosen und Wiesenkräuter – unter besonderer Berücksichtigung sekundärer Pflanzeninhaltsstoffe. Beiträge zur 13. Wissenschaftstagung Ökologischer Landbau, Hochschule für nachhaltige Entwicklung Eberswalde, Köster, Berlin, 17.–20. März 2015, https://orgprints.org/id/eprint/27259/1/27259_hamacher.pdf, (downloaded 2021).

Hatfield R., Muck R., 1999: Characterizing proteolytic inhibition in red clover silage. Proc. 12th international silage conference, Swedish University of Agricultural Sciences, Uppsala, Sweden, 147–148.

Huhman D.V., Sumner L.W., 2002: Metabolic profiling of saponins in Medicago sativa and Medicago truncatula using HPLC coupled to an electrospray ion-trap mass spectrometer. Phytochemistry, 59, 347–360.

Jansman A.J.M., Verstegen M.W.A., Huisman J., 1993: Effects of dietary inclusion of hulls of faba beans (Vicia faba L.) with a low and high content of condensed tannins on digestion and some physiological parameters in piglets. Anim. Feed sci. Tech., 43, 239–257.

Jones B.A., Muck R.E., Hatfield R.D., 1995: Red clover extracts inhibit legume proteolysis. J. Sci food agric., 67, 329–333.

Johnson I.T., Gee J.M., Price K., Curl C., Fenwick G.R., 1986: Influence of saponins on gut permeability and active nutrient transport In Vitro. J. Nutr., 116, 2270–2277.

Jurysta M., 1979: Haemolytic micromethod for rapid estimation of toxic alfalfa saponin. Acta Agrobot., 32, 5–11.

Kalac P., Price K.R., Fenwick G.R., 1996: Changes in saponin content and composition during the ensilage of alfalfa (Medicago sativa L.), Food chem., 56, 377–380.

Kofahl A., 2009: Methodische Untersuchungen zur Beurteilung der proteolytischen Aktivität, der Proteolyse und der Desmolyse bei der Silierung eiweißreicher Grünfutterleguminosen. Dissertation, Universität Rostock, 2009.

Kozłowska M., Cieślak A., Jóźwik A., El-Sherbiny M., Stochmal A., Oleszek W., Szumacher-Strabel M., 2020: The effect of total and individual alfalfa saponins on rumen methane production. J. Sci. Food agric., 100, 1922–1930.

Leamaster B.R., Cheeke P.R., 1979: Feed preferences of swine: alfalfa meal, high and low saponin alflalfa, and quinine sulfate. Can. J. Anim. Sci., 59, 467–469.

Lee M.R.F., Winters A.L., Scollan N.D., Dewhurst R.J., Theodourou M.K., Minchin F.R., 2004: Plant-mediated lipolysis and proteolysis in red clover with different polyphenol oxidase activities. J. Sci. Food agric., 84, 1639–1645.

Lee M.R.F., Parfitt L.J., Scollan N.D., Minchin F.R., 2007: Lipolysis in red clover with different polyphenol oxidase activities in the presence and absence of rumen fluid, J. Sci. Food agric., 87, 1308–1314.

Levy M., Zehavi U., Naim M., Polacheck I., 1986: An improved procedure for the isolation of medicagenic acid 3-O-beta-D-glucopyranoside from alfalfa roots and its antifungal activity on plant pathogens. J. Agric. Food chem., 34, 960–963.

Lorenz M.M., Eriksson T., Uden P., 2010: Effect of wilting, silage additive, PEG treatment and tannin content on the distribution of N between different fractions after ensiling of three different sainfoin (Onobrychis viciifolia) varieties. Grass forage sci., 65, 175–184.

Maliar T., Drobna J., Kraic J., Maliarova M., Jurovata J., 2011: Proteinase inhibition and antioxidant activity of selected forage crops. Biologia, 66, 96–103.

Massiot G., Lavaud C., Le Men-Olivier L., Van Binst G., Miller S.P.F., Fales H.M., 1988: Structural elucidation of alfalfa root saponins by mass spectrometry and nuclear magnetic resonance analysis, J. Chem. Soc., Perkin Transactions 1, 3071–3079.

Massiot G., Lavaud C., Besson V., Le Men-Olivier L., Van Binst G., 1991: Saponins from aerial parts of alfalfa (Medicago sativa), J. Agric. Food chem., 39, 78–82.

Massiot G., Lavaud C., Benkhaled M., Le Men-Olivier L., 1992: Soyasaponin VI, a new maltol conjugate from alfalfa and soybean, J. Nat. Prod., 55, 1339–1342.

Mazahery-Laghab H., Yazdi-Samadi B., Bagheri M., Bagheri A.R., 2010: Alfalfa (Medicago sativa L.) shoot saponins: identification and bio-activity by the assessment of aphid feeding. Br. J. Nutr., 105, 62–70.

Norioka N., Hara S., Ikenaka T., Abe J., 1988: Distribution of the Kunitz and the Bowman-Birk family proteinase inhibitors in leguminous seeds. Agric. Biol. Chem. 52, 5/1988, 1245–1252, DOI:10.1080/00021369.1988.10868815.

Oleszek W., Jurzysta M. 1986a: Isolation, chemical characterization and biological activity of alfalfa (Medicago media Pers.) root saponins. Acta soc. Bot. Pol., 55, 23–33.

Oleszek W., Jurzysta M., 1986b: Isolation, chemical characterization and biological activity of red clover (Trifolium pratense L.) root saponins. Acta soc. Bot. Pol., 55, 247–252.

Oleszek W., Price K.R., Colquhoun I.J., Jurzysta M., Ploszynski M., Fenwick R.G., 1990: Isolation and identification of alfalfa (Medicago sativa L.) root saponins: their activity in relation to a fungal bioassay. J. Agr. Food chem., 38, 1810–1817.

Oleszek W., Jurzysta M., Ploszynski M., Colquhoun I.J., Price K.R., Fenwick G.R., 1992: Zahnic acid tridesmoside and other dominant saponins from alfalfa (Medicago sativa L.) aerial parts, J. Agric. Food chem., 40, 191–196.

Oleszek W., Nowacka J., Gee J.M., Wortley G.M., Johnson I.T., 1994: Effects of some purified alfalfa (Medicago sativa) saponins on transmural potential difference in mammalian small intestine. J. Sci. Food agric., 65, 35–39.

Ortiz L.T., Centeno C., Trevino J., 1993: Tannins in faba bean seeds: effects on the digestion of protein and amino acids in growing chicks. Anim. Feed sci. Tech., 41, 271–278.

Pecetti L., Tava A., Romani M., De Benedetto M.G., Corsi P., 2006: Variety and environment effects on the dynamics of saponins in lucerne (Medicago sativa L.). Eur. J. Agron., 25, 187–192.

Pedersen M.W., Zimmer D.E., McAllister D.R., Anderson J.O., Wilding M.D., Taylor G.A., McGuire C.F., 1967: comparative studies of saponin of several alfalfa varieties using chemical and biochemical assays 1. Crop sci., 7, 349–352.

Peterson D.W., 1950: Some properties of a factor in alfalfa meal causing depression of growth in chicks. J. Biol. Chem., 183, 647–653.

Pleger L., Weindl P.N., Weindl P.A., Carrasco L.S., Kienzle E., Bellof G., 2018: Precaecal digestibility of alfalfa products as an organic feedstuff in broilers. In Proceedings of the 22nd congress of the European society of veterinary and comparative nutrition, September 6th–8th 2018, Munich, Germany, 40.

Pleger L., Weindl P.N., Weindl P.A., Carrasco L.S., Leitao C., Zhao M., Schade B., Aulrich K., Bellof G., 2020: Effects of increasing alfalfa (Medicago sativa) leaf levels on the fattening and slaughtering performance of organic broilers. J. Anim. Physiol. Anim. Nutr., 104, 1317–1332.

Pleger L., Weindl P.N., Weindl P.A., Carrasco L.S., Leitao C., Zhao M., Aulrich K., Bellof, G., 2021: Precaecal digestibility of crude protein and amino acids from alfalfa (Medicago sativa) and red clover (Trifolium pratense) leaves and silages in broilers. Anim. Feed sci. Technol., 275, 114856.

Ponte P.I.P., Mendes I., Quaresma M., Aguiar M.N.M., Lemos J.P.C., Ferreira L.M.A., Soares M.A.C., Alfaia C.M., Prates J.A.M., Fontes C.M.G.A., 2004: Cholesterol levels and sensory characteristics of meat from broilers consuming moderate to high levels of alfalfa. Poult. Sci., 83, 810–814.

Preissinger W., Lindermayer H., Propstmeier G., Hahn E., 2016: Luzerneheu, Grascobs und Maissilage in der Ferkelfütterung – Auswirkungen auf Futteraufnahme, Kotbeschaffenheit und Leistung. Proteinmarkt.de [Downloaded 3.1.19].

Pütz S., 2014: Entwicklung und Validierung von praxistauglichen Maßnahmen zum Verzicht des routinemäßigen Schwänzekupierens beim Schwein in der konventionellen Mast. Dissertation, Georg-August-Universität Göttingen, Fakultät für Agrarwissenschaften.

Rayudu G.V.N., Kadirvel R., Vohra P., Kratzer F.H., 1970: Toxicity of tannic acid and its metabolites for chickens. Poult. sci., 49 957–960.

Renaudeau D., Jensen S.K., Ambye-Jensen M., Adler S., Bani P., Juncker E., Stødkilde L., 2022: Nutritional values of forage-legume-based silages and protein concentrates for growing pigs. Animal, 16, 100572. doi:10.1016/j.animal.2022.100572.

Sakamoto S., Kofuji S., Kuroyanagi M., Ueno A., Sekita, S., 1992: Saponins from Trifolium repens. Phytochem., 31, 1773–1777.

Sen S., Makkar H.P.S., Becker K., 1998: Alfalfa saponins and their implication in animal nutrition. J. Agr. Food chem., 46, 131–140.

Serrano J., Puupponen-Pimiä R., Dauer A., Aura A.-M., Saura-Calixto F., 2009: Tannins: current knowledge of food sources, intake, bioavailability and biological effects, Mol. Nutr. Food res., 53 (S2), 310–S329.

Shany S., Gestetner B., Birk Y., Bondi A., 1970: Lucerne saponins III. – effect of lucerne saponins on larval growth and their detoxification by various sterols. J. Sci. Food agric., 21, 508–510.

Southan S., Wright A.J.A., Price K.R., Fairweather-Tait S.J., Fenwick G.R., 1988: The effect of three types of saponin on iron and zinc absorption from a single meal in the rat. Br. J. Nutr., 59, 389–396.

Stangl G.I., 2011: Die Nährstoffe und ihr Stoffwechsel. Editors: Roth F.X., Schwarz F.J., Stangl G.I., 2011: Tierernährung, 13. überarbeitete Auflage, DLG- Verlag, Frankfurt am Main.

Sullivan M.L., Hatfield R.D., 2006: Polyphenol oxidase and o-diphenols inhibit postharvest proteolysis in red clover and alfalfa. Crop sci., 46, 662–670.

Szumacher-Strabel M., Stochmal A., Cieslak A., Kozłowska M., Kuznicki D., Kowalczyk M., Oleszek W., 2019: Structural and quantitative changes of saponins in fresh alfalfa compared to alfalfa silage. J. Sci. Food agric., 99, 2243–2250.

Tava A., Oleszek W., Jurzysta M., Berardo N., Odoardi M., 1993: Alfalfa saponins and sapogenins: isolation and quantification in two different cultivars. Phytochem. Anal., 4, 269–274.

Tava, A., Odoardi M., Oleszek W., 1999: Seasonal changes of saponin content in five alfalfa (Medicago sativa) cultivars. Agric. Mediterr., 129, 111–116.

Tava A., Avato P., 2006: Chemical and biological activity of triterpene saponins from Medicago species. Nat. Prod. Commun., 1(12).

Van Leeumen P., Jansam A.J.M., Wiebenga J., Koninkx J., Mouwen J., 1995: Dietary effects of faba-bean (Vicia faba L.) tannins on the morphology and function of the small-intestinal mucosa of weaned pigs. Br. J. Nutr., 73, 31–39.

Vetter J., 2000: Plant cyanogenic glycosides. Toxicon, 38, 11–36.

Weltin J., Carrasco S., Berger U., Bellof G., 2014: Luzernesilage aus spezieller Nutzung und technologischer Aufbereitung in der ökologischen Geflügel- und Schweinefütterung. Endbericht BÖLN-Projekt, FKZ 11OE077. http://orgprints.org/26279/1/26279-11OE077-hswt-bellof-2014-luzernesilagetierernaehrung.pdf.

Weindl P.A., Kirn A., Bellof G., 2021: Saponine in der Luzernepflanze: Gehalte und Wirkungen beim monogastrischen Tier. Editors: Bellof G., Weindl P.A., Weindl P.N., 2021: Grünleguminosen als Eiweiß- und Raufuttermittel in der ökologischen Geflügel- und Schweinefütterung (GRÜNLEGUM). Schlussbericht BÖLN Projekt FKZ 2815OE039, 2815OE077, 2815OE079, 2815OE076, 2815OE073. https://orgprints.org/id/eprint/42739.

Wina E., Muetzel S., Becker K., 2005: The impact of saponins or saponin-containing plant materials on ruminant production a Review. J. Agr. Food chem., 53, 8093–8105.

3.4 Lucerne (*Medicago sativa L.*) and refining products of lucerne

Gerhard Bellof

3.4.1 *Whole plant of lucerne*

3.4.1.1 Ingredients

3.4.1.1.1 *Nutrients*

Table 3.4.1-1 shows the value-determining ingredients for lucerne plants (fresh) from different harvesting stages.

TABLE 3.4.1-1 Ingredients (g/kg) in lucerne (*Medicago sativa L.*), harvested at different stages

Ingredient	Lucerne		Lucerne		Lucerne		Lucerne	
	1. Cut, Bud		1. Cut, Flowering start		1. Cut, Flowering end		2. Cut, Before bud	
DM	180	1000	200	1000	210	1000	210	1000
CP	39	217	38	188	35	169	51	243
CL	6	33	6	30	6	28	7	33
CF	41	228	55	276	68	325	54	259
aNDFom	85	472	109	545	130	620	78	371
ADFom	49	272	63	315	76	360	70	334
STA	0	0	0	0	0	0	0	0
SUG	3	17	5	25	11	54	11	53
NFE	81	450	81	404	79	377	75	359
ASH	23	128	20	102	21	101	22	106

BELLOF *ET AL.* (2021); LFL BAYERN (2021)

The different vegetation stages – illustrated using the example of the first growth – lead to a characteristic shift in the most important constituents in the lucerne plant. With the entry into the generative phase, the dry matter content increases. The crude protein content falls below the 20% mark in the dry matter, while the fibre content (CF, NDF, ADF) increases (table 3.4.1-1). The growths from the 2nd cut and the subsequent cuts tend to achieve higher crude protein contents with moderate fibre contents. Lucerne, harvested at an early stage, can be described as a protein-emphasising coarse forage.

The comparatively high protein content and the low level of easily soluble carbohydrates make it difficult to ensile lucerne. The high calcium content (13–20 g/kg DM) also contributes to this. The situation is described by the ratio of sugar content (Z) to buffer capacity (PK) in the harvested crop. Lucerne has a very low Z/PK ratio (lucerne 0.9; grasses: 2.4; red clover: 1.7) (Steinhöfel, 2008). The preferred conservation method for lucerne is therefore natural drying into hay.

Table 3.4.1-2 shows the most important ingredients for lucerne and lucerne grass – after advertising as silage, hay or hot air drying. The combination of lucerne and grass can improve ensiling success due to the increased sugar supply in the crop. Wilting to at least 40% dry matter is an important prerequisite for successful ensiling, especially for lucerne.

TABLE 3.4.1-2 Ingredients (g/kg) in different silages, hay and cobs of lucerne

Ingredient	Lucerne-grass, silage		Lucerne, silage		Lucerne, hay		Lucerne, cobs	
	1. Cut, flowering		1. Cut, flowering		1. Cut, flowering		1. Cut, before bud Hot-air dried	
DM	400	1000	400	1000	880	1000	890	1000
CP	74	184	78	194	158	180	187	210
CL	12	31	15	38	18	20	28	31
CF	116	291	115	287	264	300	165	185
aNDFom	207	518	208	519	440	500	265	410
ADFom	121	302	128	320	271	308	214	240
STA	0	0	0	0	0	0	0	0
SUG	8	21	0	1	29	33	44	50
NFE	146	364	144	360	336	382	391	439
ASH	45	113	48	119	104	118	120	135

DLG (2014); LFL BAYERN (2014, 2021); MEYER AND COENEN (2014); CVB (2016); FEEDIPEDIA (2017)

Lucerne hay from natural drying has similar contents to lucerne silage for the value-determining ingredients. The crude protein content is at a similar level to the original material. The same applies to the fibre content. Natural drying is associated with high crumbling losses. Therefore, lucerne should be wilted to 50 to 60% dry matter and then finally dried under a roof or in loosely pressed round bales with ventilation. The drying process is slow due to the thick stalks. Drying with warm air is therefore advisable.

Another method of preserving lucerne plants is hot air drying. Here, the plants are preferably harvested at a very early stage of growth (before or in the bud) (table 3.4.1-2). The gentle drying process results in low losses and the ingredients remain at the level of the original material. The lucerne meal obtained in this way – usually offered in pelletised form – can be classified as a high-protein concentrated feed.

3.4.1.1.2 *Minerals and other ingredients*

Table 3.4.1-3 shows the contents of minerals (bulk elements) for lucerne whole plants (fresh) from different harvesting stages as well as lucerne silage and dried products.

Lucerne plants contain high levels of calcium (13–20 g Ca/kg dry matter). The Ca content decreases as the vegetation stage progresses. The Ca content of the starting material is largely retained in the preservation products. The phosphorus content, on the other hand, is at a medium level. The sodium content only reaches the 1 g/kg

TABLE 3.4.1-3 Content of minerals (g/kg DM) in lucerne, lucerne silage, lucerne hay and lucerne cobs

Ingredient	Lucerne	Lucerne	Lucerne, silage	Lucerne, hay	Lucerne, cobs
	1. Cut, Flowering start	2. Cut, Before bud	1. Cut, Flowering	1. Cut, Flowering	1. Cut, before Bud Hot-air dried
DM	200	210	412	880	890
Calcium	20.0	13.0	13.0	16.0	18.0
Phosphorus	2.8	3.4	3.2	3.5	3.8
Magnesia	2.7	1.9	2.2	3.1	3.0
Sodium	1.0	0.0	0.6	0.8	0.5
Potassium	26.0	31.1	30.0	30.0	27.0
Chlorine	6.7	3.0	5.5	6.0	6.0
Sulfur	4.5	2.0	2.0	3.0	2.0

DLG (2014); LFL BAYERN (2014, 2021); MEYER AND COENEN (2014); CVB (2016); FEEDIPEDIA (2017)

DM mark in the 1st growth. This is clearly missed in the subsequent cuts (table 3.4.1-3). In contrast, lucerne and its preserves contain very high levels of potassium. The chlorine and sulphur contents for the 1st cut are at a medium level; however, they fall significantly in young used growth from the 2nd cut. This is reflected in the forage cation-anion balance (DCAB). The lucerne plants shown in table 3.4.1-3 have very different DCAB values (lucerne 1st cut 239 and for lucerne 2nd cut 528 meq/kg dry matter). The trace element contents are at a similar level to the growth from permanent grassland. However, lucerne has a significantly lower iron content.

Lucerne plants have a high ß-carotene content. However, the content values decrease significantly with increasing vegetation stage. According to Nehring (1970), the ß-carotene content in the lucerne plant (1st cut) decreases from 330 mg/kg DM in the "pre-bud" stage to 170 mg/kg DM "in full bloom". The high initial ß-carotene content is almost completely present in the gently dried lucerne meal and is retained in the pelleted material or cobs even after prolonged storage. Lucerne meal can therefore be used as a source of ß-carotene.

As described in detail in section 3.3.1, lucerne has a medium to high saponin content. Ensiling leads to a conversion of the saponin fractions. In the diet of ruminants, these substances only appear to have negative effects if the daily feed intake is high (3.4.1.3).

3.4.1.2 Feed value

3.4.1.2.1 *Protein*

3.4.1.2.1.1 Ruminants

As the vegetation stage progresses, the proportion of rumen-stable protein (UDP) in the total crude protein increases in lucerne. For example, growths in the "in bud" stage have a UDP content of 15%. At the "beginning of flowering" stage, this rises to 20% UDP and reaches its maximum value at the "end of flowering" stage with a UDP content of 25% (LfL Bayern, 2021). The situation is similar for the subsequent cuts.

Due to decreasing crude protein contents over the course of the vegetation, this results in almost constant UDP contents for the different stages of lucerne.

However, as the proportion of rumen-degradable protein (RDP) dominates in lucerne, there is a surplus of nitrogen in the rumen. This must be secured by a targeted energy supply for microbial protein synthesis.

With a sufficient degree of wilting and successful lactic acid fermentation – without butyric acid formation – the outlined UDP proportions should also be achieved in lucerne silage.

Lucerne products from hot-air drying are high-quality protein feed for high-yielding ruminants. Due to the gentle drying process, an increased UDP content of 35% can be assumed (LfL Bayern, 2021). This means that such products can be used as high-quality protein feedstuffs in the feeding of high-yielding ruminants.

3.4.1.2.1.2 Monogastrics

Table 3.4.1-4 shows the contents of important amino acids for lucerne plants (fresh) from different stages of utilisation.

The lucerne variants documented in table 3.4.1-4 correspond to the variants shown in table 3.4.1-1. The amino acid contents reflect the protein contents. The contents of the amino acids lysine and threonine dominate. The sulphur-containing amino acids methionine (Met) and cysteine (Cys) are at a comparatively favourable level (2.5 g Met and Cys/100 g crude protein, 2nd cut, before budding). This is particularly true in comparison to grain legumes. Pea protein, for example, has only 2.3 g Met and Cys per 100 g crude protein. The methionine content in lucerne protein (1.4 g Met/100 g crude protein) is roughly on a par with soya protein (1.5 g Met/100 g crude protein).

The supply of protein and essential amino acids shown makes lucerne an attractive feedstuff for feeding pigs or poultry on organic farms. According to the requirements of the EU Organic Farming Regulation (2008), these animals must be given roughage on a daily basis. In addition to the labour aspect, the nutritional contribution is also relevant. Various studies have therefore tested lucerne silage for use in the feeding of growing pigs. The results of digestion trials with such silages are summarised in table 3.4.1-5.

The silages tested reached a considerable level with at least 21% CP in the dry matter. The digestibility of organic matter (OMD) varies between 37% and 55%, while the digestibility of crude protein (CPD) is between 44.1% and 56.3%. There is a close negative relationship between crude fibre content and nutrient digestibility. High crude fibre contents should therefore be avoided as far as possible by early

TABLE 3.4.1-4 Content of amino acids (g/kg) in lucerne and lucerne cobs, harvested at different stages

DM/ Amino acid	Lucerne		Lucerne		Lucerne		Lucerne, cobs	
	1. Cut, Bud		1. Cut, Flowering end		2. Cut, Before bud		1. Cut, Before bud Hot-air dried	
DM	180	1000	210	1000	210	1000	890	1000
Lysine	1.9	10.4	1.7	8.1	3.2	15.4	9.0	10.1
Methionine	0.5	2.9	0.5	2.2	0.7	3.4	2.5	2.8
Cysteine	0.4	2.4	0.4	1.9	0.6	2.7	2.1	2.3
Threonine	1.6	8.7	1.4	6.8	2.2	10.4	7.5	8.4
Tryptophan	0.6	3.2	0.5	2.5	0.6	2.7	2.8	3.1

BELLOF *ET AL.* (2021); LFL BAYERN (2021)

TABLE 3.4.1-5 Content of crude fibre (CF) and crude protein (CP); apparent total tract digestibility of organic matter (OMD) and crude protein (CPD) of lucerne silage for growing pigs – literature overview

Item		Lucerne, silage[A,1]	Lucerne, silage[B,2]	Lucerne, silage[C,3]	Lucerne, silage[D,3]	Lucerne, silage, extruded[D,3]
CF	g/kg DM	209	236	306	226	176
CP	g/kg DM	222	213	225	250	283
OMD	%	56.5	53.6	37.0	42.7	72.0
CPD	%	52.9	56.3	44.1	51.5	70.1

1 Fourth cut, chopped.
2 Third cut, chopped.
3 Second cut, chopped.
A Weber *et al.* (2024a).
B Messinger *et al.* (2021).
C Weltin *et al.* (2014).
D Berger (2012).

cutting. Mechanical processing of the silage with an extruder apparently leads to cell disintegration and thus to increased digestibility (72% OMD).

Silage from early-grown lucerne and red clover can be used in organic poultry feed. This fulfils the requirement to provide roughage and provides an additional source of methionine from local sources. Table 3.4.1-6 shows studies that have investigated the digestibility of amino acids in broilers.

While Ritteser and Grashorn (2015) only determined below-average digestibility values for an lucerne silage, Pleger *et al.* (2021a) and Pleger *et al.* (2021b) found considerably high content values for corresponding silages. In particular, the lucerne silage investigated in the 2nd study had a favourable composition (38% DM; 24% CP and 27% CF in the DM) and high digestibility values (table 3.4.1-6).

3.4.1.2.2 *Energy*

Green fodder from forage crops are typical feedstuffs for feeding ruminants. Depending on the time of cutting (table 3.4.1-1), lucerne plants have only low to medium nutrient digestibilities (56 to 62% OMD) for cattle feed and thus correspondingly low to medium levels of metabolisable energy.

Natural drying is associated with high crumbling losses. For this reason, lucerne is best wilted to 50 to 60% dry matter and then dried further under cover or in loosely pressed round bales with ventilation. Drying is slow due to the thick stalks. Drying with warm air is therefore advisable (LfL Bayern, 2013).

For horse feeding, the lucerne from the 1st cut described shows a low content of metabolisable energy (6.2 to 7.5 MJ/kg DM).

For pigs, the lucerne silage shown in table 3.4.1-5 (Weber *et al.*, 2024a) has a metabolisable energy content of 8.7 MJ/kg DM.

TABLE 3.4.1-6 Studies on apparent precaecal digestibility (pcd) of crude protein and amino acids (%) of different lucerne silages for broilers

Item	Lucerne silage[A]	Lucerne silage[B]	Lucerne silage[C]
CP	63	88	49
Lysine	60	100	33
Methionine	53	84	48
Cysteine	67	53	–
Threonine	51	78	49
Tryptophan	n.d.	77	n.d.
Leucine	78	81	n.d.
Valine	58	77	n.d.
Arginine	79	79	44

A Pleger *et al.* (2021a).
B Pleger *et al.* (2021b).
C Ritteser and Grashorn (2015).

3.4.1.3 Feeding recommendations

The use of fresh green fodder is specified for organic cattle, sheep and goat feeding. The lucerne plants mentioned in the previous sections can also be used in fresh or preserved form as roughage in organic pig and poultry feed.

3.4.1.3.1 *Cattle, sheep, goats*

Lucerne is a roughage with a high protein supply for dairy cows. It also has a very good structural effect for rumen digestion. As already described in 3.3.1 and 3.3.2, lucerne contains saponins and phytoestrogens. Saponins can trigger rumen flatulence (tympany) in cattle. An increased intake of phytoestrogens leads to changes in the oestrus behaviour of female cattle. For this reason, dairy cows should not be fed more than 10 kg dry matter per day of lucerne (Jilg, 2001).

Lucerne is only suitable for daily grazing to a limited extent. The time window for optimum utilisation – the highest possible feed value – is very narrow with lucerne. There is therefore a risk that overgrown and therefore very woody plants will be harvested. The stalks are usually spurned by the animals. There is therefore much to be said in favour of preserving lucerne in the form of silage or hay.

The use of lucerne silage in dairy cattle feed promotes the production of acetate in the rumen. In the study by Arndt *et al.* (2015), the acetate:propionate ratio increased with an increasing proportion of lucerne silage, which is underpinned by the increasing fibre digestibility. Brito and Broderick (2006) were able to demonstrate increasing milk fat content through the use of lucerne silage.

DM intake, milk yield (FCM) and milk fat content increase with increasing use of lucerne silage. Brito and Broderick (2006) demonstrated increasing NDF and ADF digestibilities through the use of lucerne silage in the ration. Due to the higher proportion of lucerne silage, the rumen pH was below 6 for a shorter period of time and the fluctuations in rumen pH were lower.

The use of lucerne silage or hay in sheep or goat feeding is only recommended to a limited extent. Due to the pronounced tendency to feed selection in these species, the positive effects described for dairy cow feeding hardly materialise.

3.4.1.3.2 *Horses*

Lucerne is also valued in horse feed as a source of protein with high digestibility. The energy content is often between that of straw and hay but, as already explained, is determined by the age of the lucerne at harvest. The low sugar content is interesting, so that lucerne hay can partly replace hay from permanent grassland if there is a corresponding dietary requirement (laminitis). The hard stalks in lucerne hay are problematic as they can lead to lesions at the entrance to the stomach and pylorus. Therefore, pre-cut products from hot air drying – cobs or pellets – are more suitable for horses. However, studies have not shown a positive influence on stomach health, which is occasionally propagated (Vondran *et al.*, 2016; 2017).

3.4.1.3.3 *Pigs, poultry*

In conventional pig and poultry feeding, lucerne hay is successfully used as an activity feed. In practice, the constant supply of fresh lucerne hay has had a positive effect on aggressive behaviour. If tail biting or feather pecking occurs, it can be used as a distraction.

Lucerne silage can be a useful component in the organic feeding of fattening pigs. This was demonstrated by Messinger *et al.* (2022) in feeding trials with fattening pigs (30 to 110 kg liveweight). The animals achieved an average daily consumption of lucerne silage of 410 g dry matter per fattening pig (20% of the total dry matter intake). At 741 g, the daily weight gain of these animals was significantly lower than that of the silage-free control group (831 g). However, at 2.29 kg/kg gain, the concentrate expenditure per kg gain was statistically lower than for the control animals (2.56 kg/kg). A third group of animals was given additional lucerne in the form of lucerne dry leaf in addition to the concentrate mixture (see 3.2). This led to significantly reduced daily weight gains (618 g) with a comparable total feed intake. The authors suspect a cumulative effect of the saponin effect (3.3.1) as the cause of the drop in performance.

Carrasco *et al.* (2016) used lucerne silage in organic laying hen feeding. Laying hens fed lucerne silage had a significantly higher omega-3 fatty acid content in their eggs. The lucerne silage also resulted in a lower cholesterol content in the egg

(Carrasco *et al.*, 2016). Lucerne silage is an accepted feedstuff for both broilers and laying hens and can be used in the following ration proportions in organic feeding: pullet rearing and laying hens 15–20%; broilers up to 30% (in each case in relation to the DM of the total ration) (Weltin *et al.*, 2014).

3.4.2 *Refining products of lucerne*

3.4.2.1 Ingredients

3.4.2.1.1 *Nutrients*

As already described in section 3.2, small-grain legumes can be used as a starting material for further technical processing with the aim of improving the feed value. In the following sections, new products from the processing of lucerne are presented. Table 3.4.2-1 below shows important ingredients for lucerne leaves and lucerne tips. These two hot-air dried lucerne products have significantly higher CP contents and reach or exceed the 30% CP mark (in the DM). At the same time, the fibre contents are significantly lower than those of comparable whole plants (table 3.4.1-1). There are only slight differences in the CP and fibre contents between the lucerne leaves and the lucerne tips. The sugar content in the tips is 2 times higher than in the leaves.

TABLE 3.4.2-1 Ingredients (g/kg) of lucerne refined products

Ingredient	Lucerne leaves		Lucerne tips		Lucerne stems		Lucerne protein pastes	
	2. Cut, Before bud, Hot-air dried		4. Cut, Bud, Hot-air dried		2. Cut, Hot-air dried		1. Cut	
DM	900	1000	926	1000	870	1000	302	1000
CP	293	325	274	296	124	142	163	538
CL	44	49	36	39	13	15	31	102
CF	147	163	139	150	325	373	2	7
aNDFom	198	220	n.d.	n.d.	461	530	48	158
ADFom	151	168	n.d.	n.d.	376	432	14	45
STA	0	0	12	13	0	0	3	9
SUG	50	56	97	105	33	38	n.d.	n.d.
NFE	311	344	387	418	332	382	59	194
ASH	107	119	90	97	77	88	48	159

BELLOF *ET AL.* (2021); WEBER *ET AL.* (2024B); RÖSSNER *ET AL.* (2023); RENAUDEAU *ET AL.* (2022)

The “lucerne protein paste” shown in table 3.4.2-1 originates from the process described in section 3.2.2 for extracting pressed juice from the lucerne plant (Renaudeau *et al.*, 2022). The subsequent processing produces a paste with a DM content of 30%. The composition of the ingredients differs significantly from the leaves or tips. For example, the crude protein content of over 50% (in the DM) is at the level of soya extraction meal (high protein). The product has a very low fibre content but a comparatively high ash content.

The remaining plant material resulting from the high-cutting process can be referred to as “stalks”. The ingredients listed in table 3.4.2-1 for this hot-air dried material differ considerably from the other products listed there. The dominant components are fibres. The ADFom content is close to that of cereal straw. At 99 g/kg DM, the ADL content is higher than for straw. The content of non-fibrous carbohydrates (NFC) is also 225 g/kg DM, which is about twice as high as in straw. The crude protein content is also significantly higher than in straw.

3.4.2.1.2 *Minerals*

Table 3.4.2-2 lists the most important bulk elements for the two protein-rich products obtained from lucerne processing. With regard to the calcium, phosphorus and magnesium contents, there are no significant shifts compared to the starting material.

There are differences for potassium. While the lucerne leaves have a similarly high potassium content as the lucerne plant, this value is significantly lower in the lucerne tips.

Results on the content of trace elements are not yet available for the two products.

TABLE 3.4.2-2 Content of minerals (g/kg DM) in lucerne leaves and lucerne tips

Ingredient	**Lucerne leaves** **2. Cut,** **Before bud,** **Hot-air dried**	**Lucerne tips** **4. Cut,** **Bud,** **Hot-air dried**
DM	1000	1000
Calcium	19.0	19.0
Phosphorus	3.7	3.1
Magnesia	2.6	3.6
Sodium	0	0.2
Potassium	27.0	14.8
Chlorine	n.d.	0.65
Sulfur	n.d.	3.85

BELLOF *ET AL.* (2021); WEBER *ET AL.* (2024B)

3.4.2.2 Feed value

3.4.2.2.1 *Protein*

3.4.2.2.1.1 Ruminants

As shown, the crude protein contents in the two lucerne products are in the range of 30% (in the DM). Proper hot air drying leads to reduced protein solubility and an increase in the UDP content compared to the fresh crop. In the Lucerne tips investigated by Rössner *et al.* (2023a), a UDP5 content of 36% was determined. This means that Lucerne tips have great potential as a high-quality protein feed in the feeding of high-yielding ruminants.

3.4.2.2.1.2 Monogastrics

Table 3.4.2-3 shows the contents of important amino acids for lucerne leaves, lucerne tips and lucerne paste.

The lucerne products documented in table 3.4.2-3 correspond to the variants shown in table 3.4.2-1. The amino acid contents reflect the protein contents. As in the lucerne plant, the contents of the amino acids lysine and threonine dominate. The sulphur-containing amino acids methionine (Met) and cysteine (Cys) are at a high level (2.9 g Met and Cys/100 g CP), particularly for the lucerne leaves, and thus above the level of the whole plant (2.5 g Met + Cys/100 g CP). The methionine content in the protein of the lucerne leaves (1.7 g Met/100 g CP) is above the level of soya protein (1.5 g Met/100 g crude protein). The lucerne tips shown in table 3.4.2-3 do not quite reach these content values (1.4 g Met/100 g CP).

The lucerne paste has very high lysine and methionine contents. Compared to the lucerne leaves (2nd cut) also documented in table 3.4.2-3, the lysine and methionine content is twice as high. The processing of the pressed juice (section 3.2.2)

TABLE 3.4.2-3 Content of amino acids (g/kg) in lucerne refined products

DM/ Amino acid	Lucerne leaves		Lucerne tips		Lucerne protein pastes	
	2. Cut, Before bud, Hot-air dried		4. Cut, Bud, Hot-air dried		1. Cut	
DM	900	1000	926	1000	302	1000
Lysine	16.7	18.5	11.9	12.8	10.8	35.7
Methionine	4.9	5.4	4.2	4.5	3.2	10.7
Cysteine	3.7	4.1	2.4	2.6	1.2	3.9
Threonine	11.3	12.5	10.6	11.4	7.8	25.7
Tryptophan	5.0	5.6	n.d.	n.d.	3.6	11.8

BELLOF *ET AL.* (2021); WEBER *ET AL.* (2024B); RENAUDEAU *ET AL.* (2022)

increases these two amino acids – in relation to 100 g protein – compared to the protein of the lucerne leaves (e.g. lysine 6.6 g versus 5.7 g/100 g CP).

TABLE 3.4.2-4 Crude protein content, apparent total tract digestibility of organic matter (OMD) and crude protein (CPD) of lucerne refined products for growing pigs – literature overview

Item		Lucerne tips, Hot-air dried[A,1]	Lucerne tips, Ensiled[A,1]	Lucerne leaves, Hot-air dried[B,2]	Lucerne leaves, Ensiled[B,2]	Lucerne protein pastes[C]
CP	g/kg DM	274	275	198	295	538
OMD	%	64.4	69.5	45.0	63.0	67.9
CPD	%	60.0	68.8	42.7	58.0	74.8

1 Fourth cut.
2 Third cut, chopped.
A Weber *et al.* (2024a); B Messinger *et al.* (2021); C Renaudeau *et al.* (2022).

The high content of essential amino acids leads to the conclusion that these lucerne products have a high potential as feed for the amino acid supply of pigs or poultry. Various studies have therefore tested such products for use in the feeding of growing pigs. The results of digestion trials for both hot air-dried and ensiled lucerne leaves and tips are summarised in table 3.4.2-4.

The tested Lucerne tips reach a considerable level with at least 27% CP in the dry matter. About two thirds of the OM are digested. Similar values are found for the crude protein (CPD). The analysed lucerne leaves show a mixed picture.

While the ensiled variant is at a comparable level to the Lucerne tips, the hot-air-dried leaves have significantly lower digestibility values. However, the crude protein content is also significantly lower than for the ensiled leaves. Despite higher CP contents and lower fibre contents, only slightly higher OMD and CPD were obtained for the lucerne paste compared to the Lucerne tips (table 3.4.2-4).

The further investigations carried out by Renaudeau *et al.* (2022) on growing pigs showed high standardised ileal digestibilities for the lucerne paste product for the essential amino acids investigated (e.g. Lys 87.8%, Met 83.9%; Thr 85.4%; Trp 88.3%).

Table 3.4.2-5 shows the results of studies on the precaecal digestibility of crude protein and amino acids for lucerne leaves in broilers and laying hens.

In an older study (Gruhn and Wiesmüller, 1990) on colostomised laying hens, lucerne leaves – obtained according to the harvesting method of Schmidt (1997),

TABLE 3.4.2-5 Studies on digestibility of crude protein (CP) and amino acids (%) of lucerne and lucerne refined products (hot-air dried) for poultry

CP/ Amino acid	Lucerne[A] Broiler[1]	Lucerne leaves[A] Broiler[1]	Lucerne leaves[B] Broiler[1]	Lucerne leaves[C] Laying hens[2]
CP	80	26	51	62
Lysine	88	46	49	76
Methionine	93	41	61	82
Cysteine	72	–	21	59
Threonine	82	43	40	n.d.
Tryptophan	n.d.	n.d.	46	n.d.
Leucine	89	50	45	n.d.
Valine	87	48	40	n.d.
Arginine	92	51	62	86

1 Precaecal digestibility.
2 Kolostomed: excrement sampling.
A Pleger *et al.* (2021a); B Pleger *et al.* (2021b); C Gruhn and Wiesemüller (1990).

see also section 3.2.2 – were examined for the precaecal digestibility of crude protein (pcdCP) and important amino acids (method: collection of the excrement). A medium level of 62% pcdCP was found. The digestibility values for the first limiting amino acids methionine and lysine were significantly higher (table 3.4.2-5).

A contradictory picture emerges for growing chickens. Pleger *et al.* (2021a) determined high digestibility values for crude protein and amino acids (pcdAA) for both hot-air-dried and ensiled lucerne plants (table 3.4.1-6). In contrast, the pcd values determined in the same test series for hot-air-dried lucerne leaves were very low. In the subsequent study (Pleger *et al.*, 2021b) with lucerne leaves, broilers showed higher levels of pcdCP. However, the pcdAA values here also remained below those of the lucerne whole plants. The authors report a low acceptance of the animals for the feed mixtures with lucerne leaves and a below-average development of these animals. These depression effects can presumably be explained by antinutritive substances, especially saponins. The results of the present study suggest the negative influence of saponins on pcdCP and pcdAA in broilers. The higher digestibility values for the ensiled lucerne leaves and whole plants suggest that the lower saponin content of the stems (Kalac *et al.*, 1996) and thus of the whole plants and, above all, the changes in the saponin compounds and content during the ensiling process (Szumacher-Strabel *et al.*, 2019) were responsible for the improved digestibility of the silage. In the ensiling trials of Szumacher-Strabel *et al.* (2019) with lucerne, structural and quantitative changes in saponins were observed. In two detected glycosides of soyasapogenol B, the concentration of one decreased and that of the other increased. In addition, seven different Medicagenic acid glycosides were detected. The concentration of five compounds from this group of substances hardly changed as a result of ensiling, while the other two disappeared almost completely. In contrast, the concentration of two of the five zanhic acid glycosides increased significantly, while that of the other three – including a zanhic acid tridesmoside – fell sharply. This compound is perceived as the most bitter saponin in human taste perception (Oleszek *et al.*, 1992). It is true that turkeys and chickens only have two to three functional bitter receptor variants. Nevertheless, these receptors have been sufficient in both bird species to protect them from poisoning by natural, bitter-tasting toxins over the course of evolution (Behrens *et al.*, 2014). Reports from the literature show that different saponins have different biological activities (section 3.3.1). The results of the saponin analyses for the hot-air dried lucerne leaves and the ensiled lucerne plants in the study by Pleger *et al.* (2021b) suggest that the levels of saponins with antinutritional activity in monogastric animals are reduced during the ensiling process.

3.4.2.2.2 *Energy*

As explained under 3.4.2.1, the protein-rich lucerne leaves and tips are protein feedstuffs. The energy supply is to be regarded as secondary. Based on the digestion tests carried out with growing pigs (table 3.4.2-4), Messinger *et al.* (2021) calculated metabolisable energy (ME) contents of 6.6 MJ/kg DM and 9.9 MJ/kg DM for hot-air dried and ensiled lucerne leaves. Weber *et al.* (2024a) state 10.1 MJ ME/kg DM for hot-air-dried lucerne tips and 10.8 MJ ME/kg DM for ensiled lucerne tips. The low ME content for the hot-air-dried lucerne leaves can be explained by the low OMD. As explained above for growing poultry, the causes of this depression could also be attributed to a negative influence of the effective saponin compounds still present in the lucerne leaves.

No digestion test results are yet available for the lucerne stems shown in table 3.4.2-1 as a by-product from the production of protein-emphasised lucerne products. However, due to the high ADFom content indicated, a low OMD and thus a low energy content – which is likely to be at the level of cereal straw – must be assumed for ruminants.

3.4.2.3 Feeding recommendations for refining lucerne products

3.4.2.3.1 *Cattle*

As already shown in the previous sections, lucerne tips can have CP contents of up to 30% (in DM), with a UDP content of 36%. This means that Lucerne tips harvested in the high cut could also be a good alternative to the usual protein feeds such as rapeseed and soya extraction meal for lactating dairy cows and thus contribute to reducing food competition. Rössner *et al.* (2023a) investigated the use of 30% hot-air-dried lucerne tips in a concentrate mixture for dairy cows in a feeding trial with high-lactation cows (race "Deutsches Fleckvieh"). With a milk yield level of 33 kg ECM per cow per day, there were no statistically reliable differences in feed intake, milk quantity and ingredients between the lucerne-free control group and the experimental group. There were also no differences in animal health. It is therefore possible to use a mixture of up to 30% Lucerne tips in the feeding of dairy cows (corresponding to 13% of DMI) without any loss of performance.

In a further feeding trial, Rössner *et al.* (2023b) tested the use of fibre-rich lucerne stems as a structural component in mixed rations for (high-yielding) dairy cows. The lucerne stems were chopped to a length of around 11 cm before hot-air drying. The most important ingredients are shown in table 3.4.2-1. The partial mixed ration with a proportion of 1 kg lucerne stems was consumed by the dairy cows in similar daily quantities as in the control group (1 kg straw) consumed. The milk yield results (average 33 kg ECM/cow and day) between the two groups showed no statistically verified differences, nor did the rumination behaviour. Hot air-dried lucerne stems are therefore suitable as a structural component in mixed rations for lactating dairy cows.

3.4.2.3.2 *Horses*

Lucerne stems could be a possible feedstuff for feeding leisure horses. Due to their high fibre content, low sugar content and moderate crude protein content (table 3.4.2-1), they optimally meet the requirements of these animals. In addition, the technical process chain – hot air drying with possible subsequent pressing – ensures a high degree of hygiene and low dust levels. Transport and feed presentation in the barn for such "stem cobs" can also be seen as an advantage. So far, however, there is no experience of practical feeding use.

3.4.2.3.3 *Pigs, poultry*

Werner and Bussemas (2021) investigated the use of lucerne leaves in the feeding of lactating sows and their piglets up to the 10th week of life with a red clover-grass silage. The trial was carried out in accordance with the requirements of organic farming. The sow feed mixtures contained 10% hot air-dried lucerne leaves (264 g CP/kg DM; 14 g Lys/kg DM; 4 g Cys/kg DM). In the preceding gestation period, all herd sows were moderately adapted with the test feeds. No statistically reliable differences in feed intake, suckling piglet losses and number of weaned piglets were found compared to the control group. The piglets were fed from the 15th day of life. They received the respective suckling feed mixtures in separate feeders from the sows. The supplementary feed intake of the piglets in the experimental group was significantly lower than that of the animals in the control group. Despite the differences in feed consumption, the daily weight gains during the entire rearing period (62nd day of life) did not differ between the feeding groups.

Weber *et al.* (2024b) tested the use of lucerne tips in a feeding trial with lactating sows – also under organic farming management conditions (see tables 3.4.2-1 and

3.4.2-3). A proportion of 20% lucerne tips was mixed into the sow feed mixture of the experimental group. The tips partially replaced soybean cake and field beans. These were the exclusive protein feedstuffs in the control mixture. The animals in the experimental group consumed lower daily feed quantities and therefore less ME and nutrients compared to the control. The sows compensated for this with increased body fat mobilisation. The piglets' growth performance was at the same level in both groups. The authors emphasise that early adaption to the lucerne tips – use as early as the gestation period – is necessary to improve acceptance.

Lauer *et al.* (2023) used lucerne tips (285 g CP/kg DM; 11.5 g Lys/kg DM; 4.4 g Met/kg DM) in the feed of weaned piglets. The rearing mixture of the test animals contained 20% lucerne tips. These partially replaced soybean meal (SES) as protein feed. The mixture of the control group contained SES as the sole protein feed. The present results of the feeding trial show that the use of dried lucerne tips in the tested mixture proportions of 20% in complete feed mixtures for weaned piglets is feasible. Initially, feed intake was reduced and weight development initially lagged behind the control group. However, there were no differences to the control group in terms of feed conversion ratio. The changeover to mixtures with lucerne tips should be made gradually in order to accustom the animals to the bitter taste of the lucerne tips.

The possible use of increasing proportions of lucerne dry leaves in feed mixtures for fattening pigs was tested in two studies. In an acceptance trial, Messinger *et al.* (2019) investigated whether increasing proportions of lucerne dry leaves in initial fattening concentrate mixtures are consumed without problems and what effects can be expected on growth performance. In the subsequent performance trial (Messinger *et al.*, 2022), different mix proportions of lucerne dry leaves – with and without additional intake of lucerne whole plant silage – were tested on the fattening performance and carcass value of fattening pigs. The results of the feeding trial show that the use of dried lucerne leaves in the tested proportions of up to 20% in complete feed mixtures for fattening pigs is feasible. No losses in feed intake, fattening performance or carcass value were recorded that could be caused by antinutritive ingredients – such as saponins – in the lucerne plant. The calculated savings potential – compared to the control group – is 28 kg of protein feed (sum of peas, soya and sunflower cake) per fattening pig.

The combination of lucerne leaves (15% or 10% mix in the supplement) and lucerne plant silage, which was also tested, led to reduced feed intake and significantly lower animal performance. A cumulative effect of the saponin impact is assumed to be the cause (see also 3.4.1.3).

The use of dried lucerne leaves in poultry feed has been investigated in various successive studies. Pleger *et al.* (2020) examined which proportions of dried Lucerne leaves can be realised in complete feed mixtures for broiler fattening in organic faming, taking into account important performance parameters. The aim was also to investigate whether a successive increase in the proportion of lucerne leaves during the 3-phase fattening period and the drying temperature of the lucerne leaf material used have an influence on the performance of the animals. The hot-air dried lucerne leaves used had the following nutrient and amino acid contents (per kg DM): 219 g CP; 13.1 g Lys and 3.6 g Met. The lucerne leaves dried at low temperature differed slightly from these values (228 g CP; 14.1 g Lys and 3.7 g Met). These leaves had higher protein solubility and lysine availability than the hot-air dried variant and resulted in slightly better biological performance.

The other results of the trial show that the use of lucerne leaves, especially the early use (5% from phase 1) and the incorporation of higher proportions (15–20%), lead to performance depression in growing broilers.

In contrast to the results mentioned above, a fattening trial with roosters from a dual-purpose genetics showed that the use of up to 20% lucerne leaves (from the 10th–16th week of life) did not decline the growth of the animals compared to a control group (Blume *et al.*, 2021). The results were confirmed for a comparable origin (use from the 12th week of life) (Krieger *et al.* (2024)). The lucerne leaves used in these trials were obtained in a stationary plant by sieving (see 3.2.2) and had a lower crude protein content. The increased crude ash content also indicates a different composition of the material than in the experiments by Pleger *et al.* (2020). It seems conceivable that the contrasting results are due to the different production processes of the lucerne products.

The identification and quantification of the biologically active saponins responsible for this is essential for the definition of tolerable limit values for lucerne leaf material in broiler feeding.

In a further feeding trial, Pleger *et al.* (2021c) investigated whether the use of dried lucerne leaves also leads to reduced feed intake and performance depression in laying hens. The pelleted complete feed mixtures were taken from the broiler feeding trial mentioned above (Pleger *et al.* 2020). By using the identical feed mixtures (17.5% CP, 0.29% Met), it was possible to directly compare whether the lucerne leaves also have comparable effects on the laying hen. The feed mixtures consisted of 100% organic components and were designed to be isoenergetic and isonitrogenous. For a needs-based calcium supply, additional shell grit was added.

Even low mixture proportions of 10% lucerne leaves, but especially higher proportions (15%, 20%) in complete feed mixtures, led to greatly reduced performance of the laying hens. Laudadio *et al.* (2014) tested a mixture proportion of 15% lucerne meal (whole plant) in rations for laying hens, but found no differences in performance compared to the lucerne-free control group.

In the study by (Weltin *et al.* 2014), the use of lucerne whole plant silage in laying hen feed resulted in acceptable performance (see 3.4.1.3). As already described in section 3.4.2.2, saponins are converted or degraded during the ensiling process. Against this background, the feeding of lucerne leaf silage to laying hens was investigated in a trial (Weindl *et al.*, 2022). The ensiled leaf material was dried at 40 °C, ground and processed in different proportions (10% and 20%) with other feed components to form a complete feed mixture for laying hens that covered their requirements. The complete feed mixtures were offered to the animals in mealy form. Compared to the control group, feeding the feed mixtures with lucerne leaf silage already led to significantly reduced feed intake and – with a slight time lag – a drop in liveweight in the second week of the trial. In contrast, the laying performance (g egg/hen/day) was maintained at a relatively high level without significant differences between the groups. During the trial, hens in the group with the highest proportion of lucerne leaf silage (20%) in the complete feed mixture and additional outdoor access developed significant comb necrosis and green discolouration of the egg yolks. Further investigations led to the conclusion that both symptoms were due to the formation of chlorophyll derivatives (pheophorbides a (PPBa)) in the lucerne leaf silage. Due to the lipophilic properties, there is a strong accumulation of these components in the fat-rich egg yolk and thus a green colouring of the yolk. PPBa is a degradation product of chlorophyll b. It is known for its photosensitising properties. The authors come to the conclusion that due to the potentially harmful effects (comb necrosis) of lucerne leaf silage, feeding it to laying hens is strongly discouraged.

A summarise of recommendations for feeding lucerne refined products in poultry and pig production is given in table 3.4.2-6.

TABLE 3.4.2-6 Recommendations for feeding lucerne refined products in poultry and pig production (content (%) in complete concentrate mixtures)

Animal group	**Lucerne leaves (hot-air dried)**		**Lucerne leaves (ensiled)**[1]		**Lucerne tips (hot-air dried)**	
		Reference		***Reference***		***Reference***
Laying hens	5	Pleger *et al.* (2021)	0 (!)	Weindl *et al.* (2022)		
Broiler	5	Pleger *et al.* 2020				
	20	Krieger *et al.* (2024)[2]				
Sows (suckling)	10	Werner a. Bussemas (2021)			20	Weber *et al.* (2024)
Piglets (weaned)	10	Werner a. Bussemas (2021)			20	Lauer *et al.* (2023)
Fattening pigs	20	Messinger *et al.* (2022)	20	Messinger *et al.* (2021)		

1 On DM-basis.
2 Dual-purpose crosse, starting lucerne feeding starting at 12th week of life.

3.4.2.3.4 *Fishes and shrimps*

Monika Weiß, Vanessa Fuchs and Sinem Zeytin

Knowledge about the use of green legumes in fish feed is still scarce, however a few feeding trials have been conducted lucerne (*Medicago sativa or Medicago falcata*) in the diets. In general, lucerne can only be incorporated into feeds up to a level of 20%. More research is needed and is currently being conducted to improve the usability of green legumes in general, e.g., as a protein concentrate in aquafeeds, and thus strengthen the sustainability of aquafeeds from regional resources. Mozambique tilapia (*Oreochromis mossambicus*) juveniles (fingerlings), profit from the inclusion of 20% lucerne leaves in their diets, showing enhanced growth rates (Yadav *et al.*, 2022). Cytoplasmic leaf protein extracted from lucerne enhanced growth compared to a fish meal (FM) diet, with best growth rates at inclusion rates up to 35% replacement of the FM (Olvera-Novoa *et al.*, 1990) (table 3.4.2-7).

Conversely, juvenile Nile tilapia (*Orechormis niloticus*), showed reduced performance at various levels (feed intake, nutrient utilization, decreased growth) already at inclusion levels of 5% lucerne (undefined species) (Ali *et al.*, 2003).

In controlled feeding experiments Yellow perch (*Perca flavescens*) were fed a protein concentrate with 52% protein extracted from lucerne at inclusion rates of 180 g/kg, replacing all fish meal in the diet. Fish accepted the diet, the filet quality did not change and survival was not impaired, but growth was reduced (Coburn *et al.*, 2021).

The inclusion of lucerne leaves into feeds for Mrigal carp (*Cirrhinus mrigala*) did not hamper growth even at the highest inclusion rates of 50% (Vhanalakar and Muley, 2015).

Japanese sea bass (*Lateolabrax japonicus*), is the only marine species tested in feeding experiments with small-grain legumes to date and it is recommended to use lucerne leaves as a feed additive at 20 g/kg feed, which significantly improves the FCR (Xu *et al.*, 2021) (table 3.4.2-7).

TABLE 3.4.2-7 Studies testing lucerne refining products in aquafeeds for various species

Lucerne species	Target species	Initial mean weight (g)	Product	Replaced resource	Inclusion rate (%)	Recommendable inclusion rate	Notes	Reference
Medicago sativa	Tilapia (*Oreochromis mossambicus*)	Fingerlings	Leaf meal	Reference diet	5, 10, 15, 20 + 25	20	Beneficial impact on the growth of fingerlings	Yadav *et al.* 2022
	Nile tilapia (*Oreochromis niloticus*)	~12	Meal	Reference diet	0, 5, 10, 15 + 20	5	Above 5% poor nutrient utilization and reduced feed intake	Ali *et al.* 2003
	Tilapia (*Oreochromis mossambicus*)	~0.3	Chloroplastic and cytoplasmic alfalfa leaf protein concentrates	FM	15, 25, 35, 45 + 55	35% cytoplasmic leaf protein	Improved growth	Olvera-Novoa *et al.* 1990
	Common carp (*Cyprinus carpio*)	~16	Protein-xanthophylls concentrate (PX)	Addition	5	5	Improved growth, stimulation of antioxidant system	Rechulicz *et al.* 2014

TABLE 3.4.2-7 Studies testing lucerne refining products in aquafeeds for various species (*cont.*)

Lucerne species	Target species	Initial mean weight (g)	Product	Replaced resource	Inclusion rate (%)	Recommendable inclusion rate	Notes	Reference
	Grass carp (*Ctenopharyngodon idella*)	~9	Meal	Reference diet	2.5, 5, 10, 15 + 20	10	Saponins and cellulose discussed as limiting factors	Zhao *et al.* 2011
	Yellow perch (*Perca flavescens*)	~22	Protein concentrate (52% protein, low fiber)	FM	18 (total FM)		Diet accepted, fillet composition not changed, reduced growth	Coburn *et al.* 2021
	Cirrhinus mrigala	~45	Leaf meal	Ricebran, groundnut oil cake	30, 40 + 50	Up to 50	No differences in performance	Vhanalakar & Muley 2015
Yellow lucerne *Medicago falcata*	Japanese seabass (*Lateolabrax japonicus*)	~11		Feed additive	0.5, 1, 1.5, 2 + 2.5	2	Best results with 2% diet: reduced FCR; alkaline phosphatase, trypsin and lipase increased	Xu *et al.* 2021

References

Ali A., Al-Asgah N., Al-Ogaily S. and Ali S., (2003). Effect of feeding different levels of alfalfa meal on the growth performance and body composition of Nile tilapia (*Oreochromis niloticus*) fingerlings. Asian fisheries science 16(1/2): 59–68.

Arndt C., Powell J.M., Aguerre M.J., Wattiaux M.A., 2015: Performance, digestion, nitrogen balance, and emission of manure ammonia, enteric methane, and carbon dioxide in lactating cows fed diets with varying alfalfa silage-to-corn silage ratios. J. Dairy Sci. 98, 418–430. doi:10.3168/jds.2014-8298.

Behrens M., Korsching S.I., Meyerhof W., 2014: Tuning properties of avian and frog bitter taste receptors dynamically fit gene repertoire sizes. Mol. Biol. Evol. 31, 3216–3227. doi:10.1093/molbev/msu254.

Bellof G., Weindl P.A., Weindl P.N., 2021: Grünleguminosen als Eiweiß- und Raufuttermittel in der ökologischen Geflügel- und Schweinefütterung (GRÜNLEGUM). BÖLN Projekt FKZ 2815OE039, 2815OE077, 2815OE079, 2815OE076, 2815OE073. Schlussbericht. https://orgprints.org/id/eprint/42739.

Berger U., 2012: Kleegrassilage aus spezieller Nutzung in der ökologischen Schweinefütterung. Masterarbeit, Studiengang Agrarmanagement, Technische Universität München, Hochschule Weihenstephan-Triesdorf. Masterarbeit, Studiengang Agrarmanagement, Technische Universität München, Hochschule Weihenstephan-Triesdorf, 2012.

Blume L., Hoischen-Taubner S., Sundrum A., 2021: Effects of alfalfa leaf mass as a part of organic feeding strategies on growth and slaughtering performance of dual-purpose roosters: a preliminary study. European poultry science 85, 1–17.

Brito A.F., Broderick G.A., 2006: Effect of varying dietary ratios of alfalfa silage to corn silage on production and nitrogen utilization in lactating dairy cows. J. Dairy sci. 89, 3924–3938. doi: 10.3168/jds.S0022-0302(06)72435-3.

Carrasco S., Wüstholz J., Bellof G., 2016: The effect of chopped, extruded and pelleted alfalfa silage on the egg quality of organic laying hens. Animal feed science and technology 219, 94–101. doi: 10.1016/j.anifeedsci.2016.06.001.

Coburn J., Wells M.S., Phelps N.B., Gaylord T.G. and Samac D.A., (2021). Acceptance of a protein concentrate from alfalfa (*Medicago sativa*) by yellow perch (*Perca flavescens*) fed a formulated diet. Fishes 6(2): 9.

CVB, 2016: Federatie Nederlandse Diervoederketen (Hrsg.): CVB Veevoedertabel 2016. http://edepot.wur.nl/379631.

DLG, 2014: DLG-Futterwerttabellen Schweine 2014. DLG-Verlag, Frankfurt a. M., 7. Auflage.

EG 889/2008: Commission regulation (EC) No 889/2008 of 5 September 2008 laying down detailed rules for the implementation of Council Regulation (EC) No 834/2007 on organic production and labeling of organic products with regard to organic production, labeling and control. Official journal of the European Union. 18.09.2008. L250. Vol. 51. ISSN 1725-2555. Consolidated text: 1.01.2021. http://data.europa.eu/eli/reg/2008/889/2020-01-07.

Feedipedia, 2017: Animal feed resources information system – INRA CIRAD AFZ and FAO. https://www.feedipedia.org.

Gruhn K., Wiesemüller W., 1990: Untersuchungen zum Futterwert von fraktioniert geernteten Luzerneblättern an Legehybriden und Broilerzuchthennen. Archives of Animal Nutrition 40, 607–617. doi: 10.1080/17450399009428410.

Jilg T., 2001: Luzernesilage in der Milchviehfütterung. Staatliche Lehr- und Versuchsanstalt für Viehhaltung und Grünlandwirtschaft Aulendorf. Versuchsbericht 5/2001.

Kalač P., Price K.R., Fenwick G.R., 1996: Changes in saponin content and composition during the ensilage of alfalfa (*Medicago sativa* L.). Food chemistry 56, 377–380. doi: 10.1016/0308-8146(95)00185-9.

Krieger M., Haider F., Klevenhusen F., Hinrichs D., 2024: Gewichtszunahme von Zweinutzungshähnen bei einer 100% Bio-Fütterung mit 20% Luzerneblattmehl. Editors: Bruder V., Röder-Dreher U., Breuer L., Herzig C., Gattinger A., 2024: Tagungsband zur 17. Wissenschaftstagung Ökologischer Landbau, Landwirtschaft und Ernährung – Transformation macht nur gemeinsam Sinn. Gießen, 5. bis 8. März 2024, https://wissenschaftstagung.de, FiBL Deutschland e.V., Frankfurt am Main, https://doi.org/10.5281/zenodo.11204339.

Laudadio V., Ceci E., Lastella N.M.B., Introna M., Tufarelli V., 2014: Low-fiber alfalfa (*Medicago sativa* L.) meal in the laying hen diet: effects on productive traits and egg quality. Poultry science 93, 1868–1874. doi:10.3382/ps.2013-03831.

Lauer P., Weindl P.A., Bellof G., 2023: Einsatz von Luzernespitzen (heißluftgetrocknet) als Eiweißträger in der Aufzucht abgesetzter Ferkel. Tagungsband, Forum angewandte Forschung in der Rinder- und Schweinefütterung, 25.–27.04.2023, Fulda.

LfL Bayern, 2013: Luzerne: Anbau – Konservierung – Verfütterung. Bayerische Landesanstalt für Landwirtschaft, Institut für Pflanzenbau und Pflanzenzüchtung.

LfL Bayern, 2014: Futterberechnungen für Schweine. 21. Auflage. Bayerische Landesanstalt für Landwirtschaft, ES-Druck Freising-Tüntenhausen.

LfL Bayern, 2021: Zielwertfutteroptimierung (Zifo2). Version 20211201.

Messinger D., Kaindl M., Weindl P.A., Bellof G., 2019: Futterwert und Einsatz von Luzernetrockenblatt als Eiweißfuttermittel in der ökologischen Schweinemast. In: Forum Angewandte Forschung in der Rinder- und Schweinefütterung. Tagungsunterlage: Beiträge der Veranstaltung vom 02. und 03. April 2019 in Fulda. Editor: Verband der Landwirtschaftskammern, Bonn, Verband der Landwirtschaftskammern, 116–119.

Messinger D., Weindl P.A., Aulrich K., Pleger L., Bellof G., 2022: Growth performance and carcass traits of growing pigs consuming different alfalfa (*Medicago sativa*) products in organic farming systems. Acta agriculturae Scandinavica, Section A – Animal Science. doi: 10.1080/09064702.2022.2141309.

Messinger D., Weindl P.A., Aulrich K., Pleger L., Weindl P.N., Bellof G., 2021: Determination of apparent total tract digestibility of nutrients of lucerne (*Medicago sativa*) and red clover (*Trifolium pratense*) products in growing pigs. Züchtungskunde, 93, 389–405.

Meyer H., Coenen M., 2014: Pferdefütterung. Enke Verlag Stuttgart. 5. Auflage.

Nehring K., 1970: Futtermitteltabellenwerk. VEB Deutscher Landwirtschaftsverlag, Berlin.

Oleszek W., Jurzysta M., Ploszynski M., Colquhoun I.J., Price K.R., Fenwick G.R., 1992: Zahnic acid tridesmoside and other dominant saponins from alfalfa (*Medicago sativa* L.) aerial parts. J. Agric. Food chem., 40, 191–196.

Olvera-Novoa M.A., Campos S.G., Sabido M.G. and Martínez Palacios C.A., 1990: The use of alfalfa leaf protein concentrates as a protein source in diets for tilapia (*Oreochromis mossambicus*). Aquaculture 90(3): 291–302.

Pleger L., Weindl P.A., Bellof G., 2021a: Verdauungsversuche mit Grünleguminosen bei wachsendem Geflügel. Editors: Bellof G., Weindl P.A., Weindl P.N., 2021: Grünleguminosen als Eiweiß- und Raufuttermittel in der ökologischen Geflügel- und Schweinefütterung (GRÜNLEGUM). Schlussbericht BÖLN Projekt FKZ 2815OE039, 2815OE077, 2815OE079, 2815OE076, 2815OE073. https://orgprints.org/id/eprint/42739.

Pleger L., Weindl P.N., Weindl P.A., Bellof G., 2021c: Unterschiedliche Luzerneprodukte in der Legehennenfütterung. Editors: Bellof G., Weindl P.A., Weindl P.N., 2021: Grünleguminosen als Eiweiß- und Raufuttermittel in der ökologischen Geflügel- und Schweinefütterung (GRÜNLEGUM). Schlussbericht BÖLN Projekt FKZ 2815OE039, 2815OE077, 2815OE079, 2815OE076, 2815OE073. https://orgprints.org/id/eprint/42739.

Pleger L., Weindl P.N., Weindl P.A., Carrasco L.S., Leitao C., Zhao M., Schade B., Aulrich K., Bellof G., 2020: Effects of increasing alfalfa (*Medicago sativa*) leaf levels on the fattening and slaughtering performance of organic broilers. Journal animal physiology and animal nutrition 104, 1317–1332. doi: 10.1111/jpn.13353.

Pleger L., Weindl P.N., Weindl P.A., Carrasco S., Leitao C., Zhao M., Aulrich K., Bellof G., 2021b: Precaecal digestibility of crude protein and amino acids from alfalfa (*Medicago sativa*) and red clover (*Trifolium pratense*) leaves and silages in broilers. Animal feed science and technology 275, 114856. doi:10.1016/j.anifeedsci.2021.114856.

Rechulicz J., Ognik K., Grela E.R., 2014: The Effect of Adding Protein-Xanthophylls Concentrate (PX) from Lucerne (*Medicago sativa*) on Growth Parameters and Redox Profile in Muscles of carp, *Cyprinus carpio (L.)*. Journal of Fisheries and Aquatic Science 14(3), 697–703. DOI: 10.4194/1303-2712-v14_3_12.

Renaudeau D., Jensen S.K., Ambye-Jensen M., Adler S., Bani P., Juncker E., Stødkilde L., 2022: Nutritional values of forage-legume-based silages and protein concentrates for growing pigs. Animal, 16, 100572. doi:10.1016/j.animal.2022.100572.

Ritteser C., Grashorn M., 2015: Bestimmung präcecaler Verdaulichkeitskoeffizienten für heimische Energiefuttermittel für die Hühnermast. Endbericht BÖLN-Projekt FKZ 2811OE070.

Rössner R., Weindl P.A., Kraus J., Maxa J., Thurner S., Bellof G., 2023a: Einsatz von Luzernestängeln in aufgewerteten Mischrationen für laktierende Milchkühe der Rasse Fleckvieh. Tagungsband, 134. VDLUFA-Kongress, 5.–8.9.2023, Freising.

Rössner R., Weindl P.A., Kraus J., Maxa J., Thurner S., Bellof G., 2023b: Einsatz von Luzernespitzen im Milchleistungsfutter für hochleistende Kühe der Rasse Fleckvieh. Tagungsband, 134. VDLUFA-Kongress, 5.–8.9.2023, Freising.

Schmidt L., 1997: Separate harvest of leaves from the stems of lucerne stands. Forage and grassland management, session 22.

Steinhöfel O., 2008: Konservierung von Futtermitteln: Editors: Jeroch H., Drochner W., Simon O.: Ernährung landwirtschaftlicher Nutztiere, Ulmer Stuttgart, 2., überarb. Aufl., 266–280.

Szumacher-Strabel M., Stochmal A., Cieslak A., Kozłowska M., Kuznicki D., Kowalczyk M., Oleszek W., 2019: Structural and quantitative changes of saponins in fresh alfalfa compared to alfalfa silage. Journal of the science of food and agriculture 99, 2243–2250.

Vhanalakar S. and Muley D., 2015: Evaluation of *Medicago sativa* leaf powder as supplementary feed for the growth of fish, *Cirrhinus mrigala*. Bioscience discovery, 6, 1–4.

Vondran S., Venner M., Coenen M., Vervuert I., 2017: Effects of alfalfa chaff on the gastric mucosa in adult horses. Pferdeheilkunde 33, 66–71.

Vondran S., Venner M., Vervuert I., 2016: Effects of two alfalfa preparations with different particle sizes on the gastric mucosa in weanlings: alfalfa chaff versus alfalfa pellets. BMC veterinary research 12, 110.

Weber M.E., Weindl P., Lippl M., Thurner S., Bellof G., 2024b: Einsatz von Luzernespitzen als Eiweißfuttermittel in der ökologischen Fütterung säugender Sauen. Tagungsband, 17. Wissenschaftstagung Ökologischer Landbau, 5.–7.3.2024, Giessen.

Weber M.E., Weindl P.A., Bellof G., 2024a: Apparent digestibility of crude nutrients from protein-rich lucerne products for growing pigs. Züchtungskunde, 96, (6), 432–444.

Weindl P.N., Weindl P.A., Schmidt E., Pleger L., Messinger D., Aulrich K., Schade B., Lehmann M., Bellof G, 2022: Photosensitization and green egg yolks in laying hens caused by the feeding of alfalfa leaf silage. British poultry science (CBPS), 63, 6. doi: 10.1080/00071668.2022.2137387.

Weltin J., Carrasco Alarcon L.S., Berger U., Bellof G., 2014: Luzernesilage aus spezieller Nutzung und technologischer Aufbereitung in der ökologischen Geflügel- und Schweinefütterung. http://orgprints.org/26279/1/26279-11OE077-hswt-bellof-2014-luzernesilage-tierernaehrung.pdf.

Werner D., Bussemas R., 2021: Grünleguminosentrockenblatt in der Sauenfütterung. Editors: Bellof G., Weindl P.A., Weindl P.N., 2021: Grünleguminosen als Eiweiß- und Raufuttermittel in der ökologischen Geflügel- und Schweinefütterung (GRÜNLEGUM). Schlussbericht BÖLN Projekt FKZ 2815OE039, 2815OE077, 2815OE079, 2815OE076, 2815OE073. https://orgprints.org/id/eprint/42739.

Xu A., Shang-Guan J., Li Z., Chen Q., 2021: Effects of dietary alfalfa *Medicago falcata* L. powder on growth performance, nonspecific immunity and digestive enzyme activity of *Lateolabrax japonicus*. Fisheries science 87(4): 559–567.

Yadav K.K., Sharma B., Sharma S., Yadav M.K., Upadhyay B., Ojha M., 2022: Effect of alfalfa (*Medicago sativa*) leaf meal on the growth of *Orecochromis mossambicus* fingerlings. Journal of experimental zoology India 25(1).

Zhao H.-L., Zhong M., Feng J., Luo, B., Zhao H.-X., 2011: Evaluation of dry clover meal as a feed protein resource in practical diets of juvenile grass carp, *Ctenopharyngodon idella*. Acta Hydrobiologica Sinica 35(3), 467–472.

3.5 Red clover (*Trifolium pratense L.*) and refining products of red clover

Gerhard Bellof

3.5.1 *Whole plant of red clover*

3.5.1.1 Ingredients

3.5.1.1.1 *Nutrients*

Table 3.5.1-1 documents the value-determining constituents for red clover (fresh) from different cuts and harvesting stages. The different vegetation stages – illustrated using the example of the first growth – lead to a characteristic shift in the most important constituents in the red clover plant. With the entry into the generative phase, the dry matter content increases. The crude protein content falls below the 20% mark in the dry matter, while the fibre content (CF, NDF, ADF) increases (table 3.4.1-1). The growths from the 2nd cut and the subsequent cuts tend to achieve higher crude protein contents. Compared to alfalfa plants, the fibre content in red clover tends to be lower. Red clover, harvested at an early stage, can be described as a protein-emphasising coarse fodder.

The comparatively high protein content and the low level of easily soluble carbohydrates make red clover difficult to ensile (see 3.4.1). The high calcium content (13–17 g/kg DM) also contributes to this. Like lucerne, red clover has a low Z/PK

TABLE 3.5.1-1 Ingredients (g/kg) in red clover (*Trifolium pratense L.*), harvested at different stages

Ingredient	Red clover		Red clover		Red clover		Red clover	
	1. Cut, Before bud		1. Cut, Flowering start		2. Cut, Before bud		2. Cut, Flowering start	
DM	150	1000	200	1000	150	1000	180	1000
CP	31	207	32	160	33	222	30	165
CL	6	40	0	0	5	31	6	33
CF	29	193	56	282	32	216	49	274
aNDFom	63	420	111	555	65	430	97	540
ADFom	37	247	65	325	38	255	57	315
STA	0	0	0	0	0	0	0	0
SUG	10	67	16	80	12	80	14	80
NFE	66	440	85	424	62	412	77	430
ASH	18	120	21	105	18	119	18	98

BELLOF *ET AL.* (2021); LFL BAYERN (2021)

TABLE 3.5.1-2 Ingredients (g/kg) in different silages and hay of red clover

Ingredient	Red clover-grass, silage		Red clover, silage		Red clover, hay	
	1. Cut, Flowering		1. Cut, Flowering		1. Cut, Flowering	
DM	350	1000	320	1000	880	1000
CP	54	154	54	170	128	146
CL	14	40	14	45	23	26
CF	105	300	83	260	251	285
aNDFom	205	585	174	545	493	560
ADFom	119	340	101	315	286	325
STA	0	0	0	0	0	0
SUG	4	10	6	20	44	50
NFE	140	401	130	405	400	454
ASH	37	105	38	120	78	89

DLG (2014); LFL BAYERN (2014, 2021); MEYER AND COENEN (2014); CVB (2016); FEEDIPEDIA (2017)

ratio (1.7). The preferred method of preservation for red clover is therefore also natural drying into hay.

Table 3.5.1-2 documents the most important ingredients for red clover and red clover grass – after commercialisation as silage or hay. The combination of red clover and grass can improve ensiling success due to the increased sugar supply in the crop. Wilting to at least 35% dry matter is an important prerequisite for ensiling success with red clover.

Red clover hay from natural drying has a similar content to the corresponding silage in terms of the value-determining ingredients at a comparable utilisation date. The crude protein content is at a similar level to the original material. The same applies to the fibre content.

3.5.1.1.2 *Minerals*

Red clover has a similar mineral content to lucerne. However, the calcium and potassium contents are somewhat lower (table 3.5.1-3). The trace element contents are at a similar level to the growths from permanent grassland.

However, red clover has a lower iron content. Like lucerne, red clover plants have a high ß-carotene content. The content values decrease with increasing vegetation stage.

TABLE 3.5.1-3 Content of minerals (g/kg DM) in red clover, red clover silage and red clover hay

Ingredient	Red clover	Red clover	Red clover, silage	Red clover, hay
	1. Cut, Flowering start	2. Cut, Before bud	1. Cut, Flowering	1. Cut, Flowering
DM	200	150	320	880
Calcium	13.0	17.0	15.0	13.0
Phosphorus	2.5	2.8	3.0	2.5
Magnesia	3.6	3.6	2.5	2.5
Sodium	0.3	1.4	1.0	0.2
Potassium	22.0	24.0	24.0	22.0
Chlorine	6.0	7.6	6.0	5.0
Sulfur	2.5	1.8	2.5	2.0

DLG (2014); LFL BAYERN (2014, 2021); MEYER AND COENEN (2014); CVB (2016); FEEDIPEDIA (2017)

3.5.1.2 Feed value

3.5.1.2.1 *Protein*

3.5.1.2.1.1 Ruminants

In contrast to the lucerne plant, the proportion of rumen-stable protein (UDP) in red clover remains constant at around 20% of the total crude protein, regardless of the vegetation stage and cutting number (LfL Bayern, 2021).

Due to decreasing crude protein contents in the course of vegetation, this results in decreasing UDP contents for the different stages of red clover. The proportion of rumen degradable protein (RDP) dominates in red clover. This results in a surplus of nitrogen in the rumen. This must be secured by a targeted energy supply for microbial protein synthesis.

With a sufficient degree of wilting and successful lactic acid fermentation – without butyric acid formation – the outlined UDP proportions are also achieved in red clover silage. Red clover hay has a slightly higher UDP content (25%) due to the drying process.

Red clover and red clover grass – fresh or preserved – are important coarse feedstuffs in organic cattle feeding.

3.5.1.2.1.2 Monogastrics

Table 3.5.1-4 shows the contents of important amino acids for red clover plants (fresh) from different harvesting stages.

TABLE 3.5.1-4 Content of amino acids (g/kg) in red clover, harvested at different stages

DM/ Amino acid	Red clover 1. Cut, Before bud		Red clover 1. Cut, Flowering start		Red clover 2. Cut, Before bud		Red clover 2. Cut, Flowering start	
DM	150	1000	200	1000	150	1000	180	1000
Lysine	1.5	10.1	1.6	7.8	1.7	11.5	1.4	9.6
Methionine	0.5	3.5	0.5	2.7	0.5	3.1	0.5	3.3
Cysteine	0.3	2.1	0.3	1.6	0.3	2.1	0.3	2.0
Threonine	1.3	8.7	1.3	6.6	1.4	9.0	1.2	8.2
Tryptophan	0.6	3.7	0.6	2.8	0.6	3.9	0.5	3.5

BELLOF *ET AL.* (2021); LFL BAYERN (2021)

The red clover varieties documented in table 3.5.1-4 correspond to the varieties shown in table 3.5.1-1. The amino acid contents reflect the protein contents. The contents of the amino acids lysine and threonine dominate. The sulphur-containing amino acids methionine (Met) and cysteine (Cys) are at a comparatively favourable level (2.3 g Met and Cys/100 g crude protein, 2nd cut, before budding). This reaches the level of pea protein (2.3 g Met and Cys per 100 g crude protein).

The supply of protein and essential amino acids shown makes red clover an attractive feedstuff for feeding pigs or poultry on organic farms. Red clover or red clover grass – fresh or conserved – can be used to fulfil the requirement to provide roughage. In addition to the animal welfare aspect, there is also a nutritional contribution.

Messinger *et al.* (2021) determined high digestibility values for red clover silage (263 g CP/kg DM; 178 gCF/kg DM) in growing pigs. The digestibility of organic matter (OMD) was 71%, the digestibility of crude protein (CPD) was 66%. This resulted in higher digestibility values than for the lucerne silage analysed in the same study (table 3.4.1-5). The reduced protein degradation due to the polyphenol oxidase present in red clover (section 3.3.2) also favours red clover silage over lucerne silage for use in pig feeding.

Pleger *et al.* (2021b) also carried out comparative studies with broilers using red clover silage and lucerne silage (table 3.4.1-6). The red clover silage (48% DM 190 g CP/kg DM; 182 g CF/kg DM; 9.8 g Lys/kg DM; 2.8 g Met/kg DM) had comparatively high precaecal digestibilities (pcd) of crude protein and amino acids (table 3.5.2-4). Thus, early advertised red clover silage also appears to be a roughage for organic poultry feeding, which both provides animal welfare and can make an effective contribution to nutrient supply.

3.5.1.2.2 *Energy*

Green fodder from forage crops are typical feedstuffs for feeding ruminants. Depending on the time of cutting (table 3.5.1-1), red clover plants have medium nutrient digestibilities (60% to 67% OMD) for cattle feed and therefore medium levels of metabolisable energy.

Red clover from the first growth described also has an average metabolisable energy content for horse feeding (7.1 to 8.6 MJ/kg DM).

For pigs, the red clover silage tested by Messinger *et al.* (2021) in the digestion trial had a metabolisable energy content of 10.3 MJ/kg dry matter.

3.5.1.3 Feeding recommendations

3.5.1.3.1 *Cattle, sheep, goats*

The use of red clover silage in ruminants can reduce the extent of lipolysis and biohydrogenation of polyunsaturated fatty acids (PUFA) in the rumen due to the polyphenol oxidase it contains (Van Ranst *et al.*, 2011) and thus ultimately promote the levels of α-linolenic acid in the muscle (Lee *et al.*, 2009). Higher PUFA levels in the milk of lactating ruminants are therefore possible (Halmemies-Beauchet-Filleau *et al.*, 2013).

Dairy cows can be fed high daily quantities of red clover in fresh or ensiled form. The high protein surplus needs to be balanced with energy-emphasised feed.

3.5.1.3.2 *Pigs*

Werner and Bussemas (2021) determined an average daily intake of red clover-grass silage (49% DM) of 5 kg FM per animal per day in the organic feeding of lactating sows. In addition, the sows consumed 6.4 kg of concentrate mixture per day.

Fattening pigs that are fed according to the specifications of organic farming methods can consume high daily quantities of green legume silage. In the study by Messinger *et al.* (2021), fattening pigs consumed proportions of 26% red clover silage and 23% lucerne silage in the dry matter of the daily ration.

However, the concentrates were presented restrictively. In a feeding regime containing high proportions of green fodder silage, the composition of the concentrate mixture should be adjusted to the nutrient supply of the respective silage and thus retain the character of a supplementary feed.

3.5.1.3.3 *Poultry*

The digestion trials with broilers carried out by Pleger *et al.* (2021b) showed that proportions of 15% red clover silage (dried) leaves were accepted without any problems. The weight development of these animals did not differ from the broilers in the control group. Only after the mixture proportions were increased to 20% red clover silage did the consumption performance decline.

3.5.2 *Refining products of red clover*

3.5.2.1 Ingredients

3.5.2.1.1 *Nutrients*

The following section presents novel products from the processing of red clover. Table 3.5.2-1 below shows important ingredients for various red clover leaves. These two hot-air-dried products have higher crude protein contents and lower crude protein contents than comparable red clover plants (table 3.5.1-1). Compared to lucerne leaves (table 3.4.2-1), the red clover leaves do not achieve their high protein content.

The "red clover protein paste" shown in table 3.5.2-1 originates from the process described in section 3.2.2 for extracting pressed juice from the red clover plant (Renaudeau *et al.*, 2022). The subsequent processing produces a paste with a dry matter content of 28%. The composition of the ingredients differs significantly from the corresponding leaves. For example, the crude protein content of over 55% (in the DM) is at the level of soya extraction meal (high protein). The product has very low fibre content and moderate ash content.Red clover paste thus appears to be of slightly higher quality than the analogue product alfalfa paste (table 3.4.2-1).

TABLE 3.5.2-1 Ingredients (g/kg) of red clover, red clover leaves and red clover protein pastes

Ingredient	Red clover leaves		Red clover leaves		Red clover protein pastes	
	2. Cut, Before bud, Hot-air dried		4. Cut, Begin flowering, Hot-air dried		1. Cut	
DM	900	1000	887	1000	280	1000
CP	212	235	236	262	156	556
CL	63	70	36	40	31	111
CF	150	167	113	125	3	12
aNDFom	329	366	n.d.	n.d.	44	156
ADFom	209	232	n.d.	n.d.	19	67
STA	27	30	60	67	1	3
SUG	53	59	44	49	n.d.	n.d.
NFE	387	430	323	359	73	260
ASH	87	98	91	101	17	61

BELLOF *ET AL.* (2021); PLEGER *ET AL.* (2021B); RENAUDEAU *ET AL.* (2022)

3.5.2.1.2 *Minerals*

Table 3.5.2-2 lists the most important bulk elements for the red clover leaves. The calcium content is lower compared to the starting material (table 3.5.1-3), phosphorus and sodium are slightly higher: there are no significant changes for magnesium and potassium.

TABLE 3.5.2-2 Content of minerals (g/kg DM) in red clover leaves

Ingredient	**Red clover leaves** **2. Cut,** **Before bud,** **Hot-air dried**	**Red clover leaves** **4. Cut,** **Begin flowering,** **Hot-air dried**
DM	900	887
Calcium	11.0	13.1
Phosphorus	3.8	3.2
Magnesia	3.0	3.7
Sodium	0.0	0.1
Potassium	25.0	31.1

BELLOF *ET AL.* (2021); PLEGER *ET AL.* (2021B)

Results on the trace element content of red clover leaves are not yet available.

3.5.2.2 Feed value

3.5.2.2.1 *Protein*

3.5.2.2.1.1 Ruminants

As already shown for the hot-air-dried lucerne products, proper hot-air drying leads to reduced protein solubility and an increase in the UDP content in the red clover leaves compared to the fresh crop. A UDP content of 35% can be assumed (LfL Bayern, 2021). This means that red clover leaves also have great potential as a high-quality protein feed in the feeding of high-yielding ruminants.

3.5.2.2.1.2 Monogastrics

Table 3.5.2-3 shows the contents of important amino acids for red clover leaves and red clover paste. The red clover products documented in table 3.5.2-3 correspond to the products shown in table 3.5.2-1. The amino acid contents reflect the protein contents. As in the red clover plant, the contents of the amino acids lysine and threonine dominate. The sulphur-containing amino acids methionine (Met) and cysteine (Cys) are at a medium level for the red clover leaves (2.5 g Met and Cys/100 g CP) and only reach the content of the whole plant protein (2.3–2.7 g Met

TABLE 3.5.2-3 Content of amino acids (g/kg) in red clover leaves and red clover protein pastes

DM/ Amino acid	Red clover leaves		Red clover leaves		Red clover protein pastes	
	2. Cut, Before bud, Hot-air dried		4. Cut, Begin flowering, Hot-air dried		1. Cut	
DM	900	1000	887	1000	280	1000
Lysine	12.1	15.0	12.8	14.2	9.7	34.8
Methionine	3.1	4.5	3.9	4.3	2.8	10.0
Cysteine	n.d.	n.d.	2.0	2.2	0.9	3.2
Threonine	n.d.	n.d.	10.9	12.1	7.6	27.1
Tryptophan	n.d.	n.d.	3.4	3.8	3.5	12.5

BELLOF *ET AL.* (2021); PLEGER *ET AL.* (2021B); RENAUDEAU *ET AL.* (2022)

and Cys/100 g CP). The methionine content in the protein of the red clover leaves (1.6–1.9 g Met/100 g CP) reaches the level of the alfalfa leaves (1.7 g Met/100 g CP).

The red clover paste has very high lysine and methionine contents. Compared to the red clover leaves (4th cut) also documented in table 3.5.2-3, the lysine content is 2.5 times higher and the methionine content is 2.3 times higher. The preparation of the pressed juice (section 3.2.2) increases the contents of these two amino acids – in relation to 100 g protein – compared to the red clover leaves (e.g. lysine 6.3 g versus 5.4 g/100 g CP).

The high contents of essential amino acids lead to the conclusion that these red clover products have a high potential as feed for the amino acid supply of pigs or poultry. In the study by Renaudeau *et al.* (2022), red clover paste was tested for use in the feeding of growing pigs. Compared to lucerne paste, red clover paste had significantly lower levels of OMD (60.8% vs. 67.9%) and CPD (64.2 vs. 74.8) despite similar CP and fibre contents. Further investigations by Renaudeau *et al.* (2022) showed the following standardised ileal digestibilities for the first-limiting essential amino acids for the product red clover paste: Lys 78.5%, Met 74.6%; Thr 77.1% and Trp 81.8%. The values were thus significantly below the level of the lucerne paste.

Table 3.5.2-4 shows the results of studies on the precaecal digestibility of crude protein and amino acids for red clover plant silage and hot air-dried red clover leaves in broiler chickens.

A differentiated picture emerges for growing chickens. Pleger *et al.* (2021b) determined high digestibility values for crude protein and amino acids (pcdAA) for ensiled red clover plants. In contrast, the pcd values determined in the same test series for hot-air-dried red clover leaves were at a comparatively low level. This

TABLE 3.5.2-4 Studies on apparent precaecal digestibility (pcd) of crude protein (CP) and amino acids (%) of red clover silage and red clover leaves for broilers

CP/ Amino acid	Red clover silage	Red clover leaves Hot-air dried
CP	70	50
Lysine	88	61
Methionine	99	73
Cysteine	85	27
Threonine	65	48
Tryptophan	97	52
Leucine	79	50
Valine	74	12
Arginine	90	78

PLEGER *ET AL.* (2021B)

shows a parallel situation to lucerne. It can also be assumed for red clover that the levels of saponins, which have an antinutritional effect in monogastric animals, are reduced during the ensiling process (3.4.2.2).

3.5.2.2.2 *Energy*

As explained under 3.5.2.1, the protein-rich red clover leaves are protein feed. As with corresponding lucerne products, the energy supply is to be regarded as subordinate. According to our own calculations, an ME content of 10.4 MJ/kg DM can be assumed for pig feeding. The calculated value for poultry feed is 6.5 MJ ME/kg DM.

3.5.2.3 Feeding recommendations

Isolated studies on the use of red clover products are only available for monogastric livestock.

3.5.2.3.1 *Pigs*

Werner and Bussemas (2021) investigated the use of hot-air-dried red clover leaves with a red clover-grass silage in the feeding of lactating sows and their piglets up to 10 weeks of age. The trial was carried out in accordance with the requirements of organic farming. The sow feed mixtures contained 10% red clover leaves (208 g CP/kg DM; 14 g Lys/kg DM; 3 g Cys/kg DM). No statistically reliable differences in feed intake, suckling piglet losses and number of weaned piglets were found compared to the control group. The piglets were fed from the 15th day of life. They received the respective suckling feed mixtures in separate feeders from the sows.

The supplementary feed intake of the piglets from the experimental group did not differ significantly from the control group. There were also no differences in daily weight gain during the entire rearing period (62nd day of life).

3.5.2.3.2 *Poultry*

The digestion trials with broilers carried out by Pleger *et al.* (2021b) showed that 15% red clover leaves are accepted without problems in concentrate mixtures. The weight development of these animals did not differ from the broilers in the control group. Only after the proportion of red clover leaves in the mixture was increased to 20% feed intake declined.

3.5.2.3.3 *Fishes and shrimps*

Monika Weiß, Vanessa Fuchs and Sinem Zeytin

Red clover (*Trifolium pratense*) extract (Menoflavon® produced by Melbrosin (Vienna, Austria)), is rich in isoflavones. Isoflavones are secondary plant substances and are found in particular in soybeans and red clover. They have a chemical structure similar to estrogens (3.3.1). It has been suggested that phytoestrogens stimulate growth, as has been shown in other fish species (Malison *et al.*, 1988, Trudeau *et al.*, 1992). Isoflavones are also known to protect against free radicals and strengthen the immune system. The clover extract was used as a feed additive and low inclusion levels of 100 mg/kg were shown to improve growth rate, feed conversion and protein utilization significantly in *Oreochromis aureus*.

In Chinese blunt snout bream (*Megalobrama amblycephala*), the inclusion of *Trifolium sp.* had a positive impact on palatability, increased the feed intake and led to a better filet quality in terms of protein and fat content whilst maintaining the same growth performance indicators as for control diets (Jia *et al.*, 1991) (table 3.5.2-5). Considering these very positive results it is all the more regrettable that the authors used the terms "alfalfa" and "Trifolium" for the plant used in this study without defining the exact species.

Most of the studies conducted on green legume inclusion in aquafeeds so far did not focus on a balanced amino acid profile in the feeds, but instead tested upper levels of possible inclusion rates, without a precision nutrition approach.

Interest is growing in using a very promising "grass protein", which is a protein concentrate extracted from clover-grass, for formulation in aquaculture diets. These products are currently being evaluated in a series of experiments. The only product currently commercially available comes from Denmark. It is based on red and white clover as well as on grass pastures (retailer: Vestjyllands Andel, Denmark). It has high contents of CP (48–62%) and CL (9–12%).

TABLE 3.5.2-5 Studies testing red clover refining products in aquafeeds for various species

Species	Initial mean weight (g)	Product	Replaced resource	Inclusion rate (%)	Recommendable inclusion rate	Notes	Reference
Tilapia (*Oreochromis aureus*)	~0.3	(Menoflavon®) powdered red clover extract	Feed additive	50, 100, 200 mg/kg diet	100 mg/kg	Growth rate, food conversion ratio, protein efficiency ratio, and apparent net protein utilization significantly highest in groups fed the 100 mg red clover/kg diet	Turan (2006)
Chinese blunt snout bream (*Megalobrama amblycephala*)	~7	Meal	FM and soybean cake	20, 40 + 60	40	Inclusion at 20–40% levels resulted in better palatability, increased feed intake and a better quality in terms of relative carcass protein and fat content. Confusion between Trifolium and alfalfa	Jia *et al.* (1991)
African catfish (*Clarias gariepinus*)	~42	(Menoflavon®) powdered red clover extract	Feed additive	25, 50 + 75 mg/kg	75 mg/kg	Improves growth	Turan & Akyurt (2005)

References

Bellof G., Weindl P.A., Weindl P.N., 2021: Grünleguminosen als Eiweiß- und Raufuttermittel in der ökologischen Geflügel- und Schweinefütterung (GRÜNLEGUM). Schlussbericht BÖLN Projekt FKZ 2815OE039, 2815OE077, 2815OE079, 2815OE076, 2815OE073. https://orgprints.org/id/eprint/42739.

CVB, 2016: Federatie Nederlandse Diervoederketen (Hrsg.): CVB Veevoedertabel 2016. http://edepot.wur.nl/379631.

DLG, 2014: DLG-Futterwerttabellen Schweine 2014. DLG-Verlag, Frankfurt a. M., 7. Auflage.

Feedipedia, 2017: Animal feed resources information system – INRA CIRAD AFZ and FAO. https://www.feedipedia.org.

Halmemies-Beauchet-Filleau A., Vanhatalo A., Toivonen V., Heikkilä T., Lee M.R.F., Shingfield K.J., 2013: Effect of replacing grass silage with red clover silage on ruminal lipid metabolism in lactating cows fed diets containing a 60:40 forageto-concentrate ratio. J. Dairy sci. 96, 5882–5900.

Jia L., He X. and Yang Y., 1991: Evaluation of partial replacement of fish meal and soybean meal cake by alfalfa, Trifolium sp., in practical diets for Chinese blunt snout bream, Megalobrama amblycephala, fingerlings. Fish nutrition research in Asia: 119–123.

LfL Bayern, 2014: Futterberechnungen für Schweine. 21. Auflage. Bayerische Landesanstalt für Landwirtschaft, ES-Druck Freising-Tüntenhausen.

LfL Bayern, 2021: Zielwertfutteroptimierung (Zifo2). Version 20211201.

Malison J.A., Kayes T.B., Wentworth B.C. and Amundson C.H., 1988: Growth and feeding responses of male versus female yellow perch (Perca flavescens) treated with estradiol-17β. Canadian Journal of fisheries and aquatic sciences 45(11): 1942–1948.

Messinger D., Weindl P.A., Aulrich K., Pleger L., Weindl P.N., Bellof G., 2021: Determination of apparent total tract digestibility of nutrients of lucerne (Medicago sativa) and red clover (Trifolium pratense) products in growing pigs. Züchtungskunde, 93, 389–405.

Meyer H., Coenen M., 2014: Pferdefütterung. Enke Verlag Stuttgart. 5. Auflage.

Pleger L., Weindl P.N., Weindl P.A., Carrasco S., Leitao C., Zhao M., Aulrich K., Bellof G., 2021b: Precaecal digestibility of crude protein and amino acids from alfalfa (Medicago sativa) and red clover (Trifolium pratense) leaves and silages in broilers. Animal Feed Science and Technology 275, 114856. doi: 10.1016/j.anifeedsci.2021.114856.

Renaudeau D., Jensen S.K., Ambye-Jensen M., Adler S., Bani P., Juncker E., Stødkilde L., 2022: Nutritional values of forage-legume-based silages and protein concentrates for growing pigs. Animal, 16, 100572. doi: 10.1016/j.animal.2022.100572.

Trudeau V.L., Somoza G.M., Nahorniak C.S. and Peter R.E., 1992: Interactions of estradiol with gonadotropin-releasing hormone and thyrotropin-releasing hormone in the control of growth hormone secretion in the goldfish. Neuroendocrinology 56(4): 483–490.

Turan F., 2006: Improvement of growth performance in tilapia (*Oreochromis aureus Linnaeus*) by supplementation of red clover (*Trifolium pratense*) in diets. Israeli Journal of Aquaculture – Bamidgeh 58(1):34–38. DOI: 10.46989/001c.20427.

Turan F., Akyurt I., 2005: Effects of red clover extract on growth performance and body composition of African catfish *Clarias gariepinus*. Fisheries Science 71(3):618–620. DOI: 10.1111/j.1444-2906.2005.01006.x.

Van Ranst G., Lee M.R.F., Fievez V., 2011: Red clover polyphenol oxidase and lipid metabolism. Animal: an international journal of animal bioscience 5, 512–521.

Werner D., Bussemas R., 2021: Grünleguminosentrockenblatt in der Sauenfütterung. Editors: Bellof G., Weindl P.A., Weindl P.N., 2021: Grünleguminosen als Eiweiß- und Raufuttermittel in der ökologischen Geflügel- und Schweinefütterung (GRÜNLEGUM). Schlussbericht BÖLN Projekt FKZ 2815OE039, 2815OE077, 2815OE079, 2815OE076, 2815OE073. https://orgprints.org/id/eprint/42739.

Index

www.ingramcontent.com/pod-product-compliance
Lightning Source LLC
Chambersburg PA
CBHW071324230926
51849CB00002B/306